PANDA FANDOM

This book provides the first exploration of panda fandom, a term denoting the community built upon the deep and intimate emotional connections between fans and giant pandas. The book uncovers the dynamic relationship among China, the West, and the giant panda, rooted in over two centuries of cross-cultural encounters.

Drawing from extensive fieldwork at the Chengdu Research Base of Giant Panda Breeding, the book explores how panda fandom has evolved as a result of China's participation in global capitalism, cultural exchange, and diplomatic strategy. This book aims at showing that panda fandom, although taking its shape in the Chinese nation currently, has been a consequence to the expansion of Western imperialism, colonialism, and capitalist industrialization in the Chinese community since the 18th century. It undertakes the task of inquiring how the Chinese communities, through the figure of the giant panda, have responded and made sense of this modern transformation. The book is loosely organized around the chronological order of the development of the panda fandom and can be divided into two parts. The first part explains the emergence of the panda fandom, and the second part deals with interviews. Through a rich blend of personal narratives, historical documents, daily observations, and carefully crafted interviews, it provides a fresh perspective on how cuteness, biopower, and nationalism shape both Chinese identity and the lived experiences of these fans. Based on an intimate look at the bond between panda fans and their beloved animal, the book also relates the panda fandom to theories of the more-than-human and multispecies world.

The book will be beneficial to students and academicians interested in fan studies, popular culture, animal–human relations, conservation, and the consumption of animals. It will also be of use to anyone who wants to identify himself/herself as a panda fan and has navigated the panda–human relationship by their own means.

Yulei Guo is a research associate specializing in giant panda tourism at the Chengdu Research Base of Giant Panda Breeding in China. With a PhD in tourism studies from Otago University, New Zealand, and a master's degree in service management from Lund University, Sweden, she brings a global perspective to her work. Yulei also completed a visiting fellowship with the Biodiverse Anthropocene program at Oulu University, Finland, in 2024. Her research looks into animal–human interactions and the vibrant world of panda fandom, offering fresh, interdisciplinary insights into this fascinating field.

PANDA FANDOM

Nation, Capitalism, and Biopower

Yulei Guo

Routledge
Taylor & Francis Group
LONDON AND NEW YORK

Designed cover image: gettyimages.ca/VII-photo

First published 2026
by Routledge
4 Park Square, Milton Park, Abingdon, Oxon OX14 4RN

and by Routledge
605 Third Avenue, New York, NY 10158

Routledge is an imprint of the Taylor & Francis Group, an informa business

British Library Cataloguing-in-Publication Data
A catalogue record for this book is available from the British Library

Library of Congress Cataloging-in-Publication Data
Names: Guo, Yulei, author.
Title: Panda fandom / Yulei Guo.
Description: First edition. | New York : Routledge, 2025. | Includes bibliographical references and index.
Identifiers: LCCN 2024040758 (print) | LCCN 2024040759 (ebook) | ISBN 9781032841038 (hbk) | ISBN 9781032841045 (pbk) | ISBN 9781003511229 (ebk)
Subjects: LCSH: Giant panda—Social aspects. | Giant panda—Political aspects. | Human-animal relationships. | China—Foreign relations—20th century.
Classification: LCC QL795.P18 G86 2025 (print) | LCC QL795.P18 (ebook) | DDC 599.789—dc23/eng/20250206
LC record available at https://lccn.loc.gov/2024040758
LC ebook record available at https://lccn.loc.gov/2024040759

ISBN: 978-1-032-84103-8 (hbk)
ISBN: 978-1-032-84104-5 (pbk)
ISBN: 978-1-003-51122-9 (ebk)

DOI: 10.4324/9781003511229

Typeset in Times New Roman
by Apex CoVantage, LLC

For panda fans and pandas

CONTENTS

TABLES

ACKNOWLEDGMENTS

Many people, pandas, and places have come together to make this book possible. As I journeyed through the panda fandom, writing this book became a way to celebrate my moments with them all.

My first thanks go to the Panda Base, where this journey began, marking the intersection of Chinese and Western values, mentalities, and philosophies. My research career at the Panda Base offered me an invaluable opportunity to learn, appreciate, and understand Chinese culture—something I had grown estranged from due to my extended stays in Sweden and New Zealand. This distance from my native culture and language became the source of my inspiration and craft. I am captivated by the idea of falling in love with this mysterious, beautiful, and sometimes baffling culture for a second time. I am forever grateful to the Panda Base for seeing my struggles and tolerating my many adventurous and unruly experiments as I sought to make sense of this world. In this way, Chinese wisdom and culture have illuminated my research path and offered their support.

"Science with Arctic Attitude" is a gift I received from Oulu University, Finland, where this writing journey took shape. It is my absolute honor to be a visiting fellow of the Biodiverse Anthropocene program, which allowed me to reflect on panda fandom across continents, cultures, scientific disciplines, arts, species, and peoples. The Arctic attitude is about more than just survival in extreme weather and remote locations—it's about experiencing some of the most diverse and colorful pleasures I could ever imagine and applying this mindset to science. Despite its physical distance, Oulu turned out to be an ideal place to contemplate panda fandom. The Arctic spirit urged me to carefully evaluate, adapt, and take risks in my panda experiences in Chengdu, a place warmer, more humid, and louder than Oulu, giving the giant panda I carried in my mind a futuristic, crisp, and transparent ice shell.

My sincere gratitude goes to Marko Mutanen and Roger Norum, who granted me the fellowship and provided many intellectual inspirations through conferences, seminars, and lunch conversations. Alberto Amore showed me the Oulu campus and has been an excellent collaborator on our panda projects. All Anthropocene Biodiverse members hold a unique place in my memory.

David Fennell has supported and encouraged me to think and write as boldly as I can. He always offers the warmest encouragement and the most insightful comments.

My mother, Min Luo, and my father, Qiang Guo, have given me their best love and have always been fans of my research and dreams. They have also become loyal panda fans since I began my work at the Panda Base.

And, of course, the panda fans and pandas are the joy of this book. I am your fan.

学术声明

本书的所有内容均基于学术研究和分析，所讨论的观点和结论仅代表个人的学术立场，并不旨在表达或暗示任何对成都大熊猫繁育研究基地（以下简称“熊猫基地”）、大熊猫形象或中华人民共和国国家形象的贬损、冒犯或有害言论。本书始终尊重熊猫基地的形象、中华人民共和国的国家形象及大熊猫的文化地位，绝无任何贬损或不尊重的意图。

本书仅用于学术讨论和分析，未涉及任何政治立场或对特定机构、个人或国家的评价。所有材料和观点均从学术研究的角度出发，旨在客观探讨相关现象背后的文化、社会和经济因素。本书所涉及的所有内容，均为客观的学术探索，旨在探讨相关现象背后的文化、社会和经济因素，且这些研究均基于公开的学术资料和研究成果。

书中研究内容和方法符合《国务院关于印发涉及人的生命科学和医学研究伦理审查办法的通知》第三十二条的相关要求，确保研究中涉及的采访数据匿名化处理，个人经历和观察数据未干扰公共行为或涉及隐私。

作为一项跨学科的研究，本书旨在推动生物多样性保护，通过结合社会学、文化研究和生态保护等多学科方法，提升公众对生物多样性保护的认知与参与度，响应国家关于科学普及的政策目标。本书体现了国际化的研究视野和学术合作价值，得益于在芬兰奥卢大学访学期间的学术资源支持。

ACADEMIC DISCLAIMER

The content of this book is based on academic research and analysis. The views and conclusions discussed represent only the author's academic standpoint and are not intended to express or imply any disparagement, offense, or harmful remarks toward the Chengdu Research Base of Giant Panda Breeding (hereinafter referred to as "the Panda Base"), the image of the giant panda, or the national image of the People's Republic of China. This book always respects the image of the Panda Base, the national image of the People's Republic of China, and the cultural status of the giant panda, with no intention of disrespect or disparagement.

This book is intended solely for academic discussion and analysis and does not involve any political stance or evaluation of specific institutions, individuals, or nations. All materials and viewpoints are presented from an academic research perspective, aiming to objectively explore the cultural, social, and economic factors behind the related phenomena. The content discussed in this book is an objective academic exploration, aiming to investigate the cultural, social, and economic factors behind these phenomena, and all research is based on publicly available academic materials and findings.

The research content and methods of this book comply with the relevant requirements of Article 32 of the "Regulations on the Ethical Review of Life Science and Medical Research Involving Humans" issued by the State Council in 2023, ensuring that interview data used in the research is anonymized and that personal experiences and observational data do not interfere with public behavior or involve privacy.

As an interdisciplinary study, this book aims to promote biodiversity conservation by integrating methods from sociology, cultural studies, and ecological protection;

enhancing public awareness and participation in biodiversity conservation; and responding to national policy goals related to scientific popularization. This book reflects an international research perspective and academic cooperation value, benefiting from academic resources during the author's academic exchange at Oulu University in Finland.

INTRODUCTION

My first experience attending the reintroduction weekly meeting at the Chengdu Research Base of Giant Panda Breeding (hereafter referred to as Panda Base) at the end of 2018 was thrilling – I was about to learn about and witness one of the rarest and most beautiful animals on this planet! Embarking on my career at the Panda Base meant a long-awaited opportunity to get to know the giant panda. However, during the meeting, as we watched a video and discussed Qian Qian, a selected giant panda being trained in a lush forest, and as an anecdote about panda fans was casually mentioned, my attention was soon drawn to the latter. Qian Qian was a beautiful and intelligent panda, perhaps one of the most striking breed at Panda Base, selected for the training program due to her intelligence, as my colleague explained. Yet, in that moment, even Qian Qian could not compete with my curiosity about the panda fans.

Panda fans were described as passionate and devoted individuals who closely followed the pandas' every move. They inspected the pandas with high-end cameras and became upset if a panda lost a patch of hair near its nose. Discontent among panda fans often led to complaints about Panda Base, which had gradually become accustomed to their incessant intrusions. To avoid fans' surveillance, Panda Base adopted strategies such as transporting pandas in the middle of the night and establishing hotlines to handle fans' inquiries. Despite their interest in the welfare and breeding programs of the pandas, panda fans were said to serve their "co-piloting" role poorly. Their ideas were often biased and ungrounded, as they lacked scientific knowledge and professional training. Additionally, with their chatty, irrational, and irritable temperaments, fans were not seen as the ideal allies by Panda Base. More often than not, panda fans were regarded as an obsessive community that proclaimed unrealistic care and welfare agendas for the pandas. These fans were often

DOI: 10.4324/9781003511229-1

compared to Western pro-animal activists, stereotyped as extremists or "terrorists" by the scientific community and the media (Emel & Wolch, 1998).

My interest in panda fans, labeled "dangerous" and "crazy" commits, did not develop further initially since, as an employee at the Panda Base, the rule was to avoid personal contact with panda fans – the long interactions with the panda fans showed that a marketing department was needed to properly take care of the communication between the organization and fans to avoid confusions and information disparity. However, the fast growing panda fan community, driven by the popularity of mobile devices in China (Yin et al., 2014), the exponential increase of social media networking (Wang, 2016), and panda's rising Internet influence (Huang & Wang, 2019), has gradually turned the table around – panda fans were literarily every tourist entering the Panda Base, one of the perhaps most visited animal care organizations in the world boosting more than 9 million visitors since 2019. On the Internet, panda videos and clips were often among the top-trending topics on Chinese social media platforms such as the Weibo, Little Red Book, and Douyin. In 2023, Hua Hua, a 3-year-old panda bear with her distinguishable pattern and appearance, was nicked named the "top female celebrity in panda kingdom" by netizens. Hua Hua attracted as many as 264,000 visitors to Chengdu in the May Golden holiday in China in 2023. In 2024, Hua Hua was appointed as the honorary director of the Culture and Tourism Department in Chengdu. From a niche group of animal welfare advocates to a dominant segment of tourists and online communities, engaging with and understanding panda fans have become an integral routine of Panda Base. Indeed, becoming a panda fan has become a new norm.

However, the fan-ship with the giant panda appears to have been an inevitable aspect of the relationship between this animal and humans since her discovery in 1869. When the French missionary Armand David encountered the giant panda in Dengchigou, Sichuan, he described her in his scientific journal as the most beautiful and remarkable animal he had ever seen. David soon procured panda skins from locals at high prices and shipped them back to the Natural History Museum in Paris. Enthralled by his discovery, David wrote to his contact at the museum, suggesting he might have found a new species. This encounter with the giant panda seemed to cast a spell over Western biologists, who willingly devoted their time, money, and energy to the study and collection of this animal – true panda enthusiasts. As the panda made its way into the Western world, particularly in zoos around the globe, it effortlessly garnered millions of admirers, solidifying her status as an animal superstar.

The first giant panda, Su-Lin, hosted at the Brooklyn Zoo in the United States, was a sensation, drawing more than 300,000 visitors to the Brookfield Zoo within 6 months of her debut (Heller, 2015). The zoo itself claimed the acquisition of the giant panda was "the most important single achievement in collecting animals in modern times" (Croke, 2009). In 1938, four giant pandas, delivered by Floyd Tangier Smith, a Japanese-born American, arrived at the London Zoo and were visited by thousands (Nicholls, 2013). Ming, one of the four pandas, became so popular that her image was reproduced on posters, cartoons, postcards, stuffed animals, clothing, and television broadcasts (Nicholls, 2013). Chiang Yee, a Chinese poet, author, and artist, observed that crowds, especially children, gathered around Ming's enclosure to pet or cuddle the animal (Nicholls, 2013). The arrival of Chi-Chi at Berlin Zoo in 1958 attracted about 400,000 visitors to Tierpark Berlin (Berlin Zoo, 2022). The first pair of giant pandas at Ueno Zoo, Kang Kang and Lan Lan, arrived in Japan in October 1972 and drew a record 7,640,000 visitors in 1974 (Pallas, 2017). Wang Wang and Funi, who arrived at Adelaide Zoo in Australia in 2009, saw their first year draw 595,000 visitors, a 70% increase compared to the previous year (Peddie, 2010). Barua (2020) writes that the World Wildlife Foundation's use of the giant panda logo is among the most valuable trademarks of the last century.

Terms like "Pandamonium" (Chinoy, 1975), "Pandamania" (Angier, 2006; Roberts, 1988), "Panda Craze" (Barua, 2020; Cohen, 2010), and "Panda Frenzy" (VCG, 2024) have been used by Western media and the scientific community to describe the public sensation created by the panda. For example, Chinoy (1975, p. 15) coined the term "Pandamonium" to capture the attraction of Ling-Ling and Hsing-Hsing at the National Zoo in Washington, describing how the pair's playful antics "charmed an estimated 12 million people". Panda fandom continues to explore this public fascination with the giant panda but differs from these terms in two ways. First, panda fandom focuses on panda fans – those who have received less attention in the aforementioned terms. When I first heard the term "panda fan" at Panda Base, it was still a technical term used by panda biologists to describe what might be called "panda activists" in English. However, as the panda fandom grew, the term "panda fan" became more widespread, encompassing anyone in China who developed a passion for the panda. It is a term that goes beyond the panda's public appeal based on zoo visits, as captured by "Pandamonium" and "Pandamania", and addresses a national population whose fascination with the animal is the result of ongoing Western–Eastern interactions. While it might be intriguing to investigate whether the Pandamonium at the National Zoo in Washington inspired equivalent "panda manics" who caused trouble for the institution, I intend to use the terms "panda fandom" and "panda fans" to capture a specific history and materiality rooted in contemporary China – where both pandas and panda fans live and learn to "fan" each other.

When I first presented my idea of panda fandom to my colleagues in Europe, an American scholar told me he liked the term, even though he had never heard of it before. A few days later, another American colleague excitedly informed me that she had spotted panda fans on TikTok and believed there were many American "panda fans". "You know", she said, "they did exactly what you said about the Chinese panda fans". The best way to understand the Chinese specificity of panda fandom, I believe, can be through the term "sheng nü" (剩女). Literally translated as "leftover women", sheng nü was originally coined by a fashion magazine in China in 2006 to refer to single women over the age of 27 who remained unmarried, often with a derogatory connotation. In the years that followed, several studies on sheng nü explored how Chinese women navigate traditional gender roles through their modern lifestyles (Feldshuh, 2018; Ji, 2015; To, 2013). When I discussed the phenomenon of "leftover women" during my PhD years, I was informed that the concept was not unique to China – terms like "spinsters" and "pickled women" were used in other cultures as well. While "leftover women" remains a term bound to its linguistic and cultural origin, it seems to me that "panda fan" has become a more readily accepted term among my Western colleagues to describe both social media participants and their own interest in the giant panda. While the panda fans discussed in this book are all ethnically Chinese, as I conducted my fieldwork at Panda Base in 2023, and the Chinese perspective this book aims to provide is central, the term's potential international appeal or relevance should be acknowledged at the outset.

Tracing the origins of panda fandom to its Chinese audiences and communities also means that this book aims to provide a non-Western perspective on the growing interest in interwoven naturecultures (Haraway, 2013), multinaturalism (Lorimer, 2015), more-than-human worlds (Abram, 2012), posthumanism (Braidotti, 2013), multispecies (Braidotti, 2013), and nonhuman agency (Painter, 2016) in contemporary society. According to Abram (2012), human existence in the more-than-human world is in crisis, as language and technological advancements have fundamentally altered how people "make sense" of the world, to the extent that the reciprocity and embodied perception linking humans to the more-than-human world are jeopardized. Resonating with Abram's philosophy is the devoted engagement and commitment of panda fans to a species of their choice. This book is inspired by a desire to understand this loving attachment to a more-than-human species and how it pushes the boundaries of current human–animal relations. I ask whether panda fandom can be a space where fans have created a different, perhaps less violent and more caring, relationship with animals, and how this understanding can help us imagine more compassionate and inclusive more-than-human worlds.

However, this panda–human relationship is not separate from the existing social and human relations that have provided both the material foundation and means for the construction of panda fandom. While these social relations enable fans to make sense of their connections with pandas in specific ways, they have also imposed constraints on how fans can engage with pandas. This book is dedicated to acknowledging four social relations that have shaped the form, depth, and experience of panda fandom. Figuratively, the chapters are organized around these four critical paths: The nation of China, the panda as a commodified means of production, the aesthetic of cuteness, and the implementation of biopower. These pathways allow four interconnected aspects of panda fandom to emerge: The ongoing Western–Chinese interactions that have both initiated and sustained panda fandom, the panda as national property since its discovery, the cuteness aesthetic that naturalizes and exploits the panda, and biopower, which intersects with both the Chinese nation and the panda. I engage extensively with Michel Foucault's concept of biopower in this book, demonstrating how his understanding of biopower helps explain the relationship among the Chinese nation, the giant panda, the aesthetic of cuteness, and capitalist relations. I explore how biopower has been implemented in the construction of Chinese subjectivity, the regulation of the Chinese population through the idea of the nation, the anthropocentric and embodied aesthetic that bonds the giant panda to the human species, and how biopower intersects with capitalist modes of production. Three theoretical frameworks – Rey Chow's explanation of Chinese modernity, Karl Marx's theory of capitalism, and Foucault's concept of biopower – serve as the pillars around which these themes are investigated.

The use of Western theories to understand what I have specifically contextualized as Chinese is not an attempt to culturally appropriate panda fandom into Western theory or epistemology. Rather, this book aims to show that while panda fandom currently takes shape within the Chinese nation, it is a consequence of the expansion of Western imperialism, colonialism, and capitalist industrialization within the Chinese community since the 18th century. As a result, panda fandom belongs to a contemporary history in which non-Western cultures have been and continue to be modernized, Westernized, capitalized, and industrialized globally. This book seeks to explore how Chinese communities, through the figure of the giant panda, have responded to and made sense of this modern transformation. To this end, I aim to show that since the discovery of the giant panda in 1869, the Chinese community has engaged in the construction of the giant panda fandom in ways that have, by Western and modern orders, forced them to assume the role of the "other" – the Chinese, the non-Western – alongside the giant panda. Recognizing the Chinese community's participation in the construction of what I call the "giant panda industry", circulating globally, challenges Western scholars' views (Hartig, 2013; Nicholls, 2012; e.g., Songster, 2018) on the Chinese early indifference to the panda industry, the nation's political use of panda diplomacy, and the Chinese identification with the panda. These ideas must be reconsidered in light of the power relations between the West and China. Thus, the panda fandom, fully realized in China today, materializes this history and is the product of interconnected and multiple structures, such as capitalism, Orientalism, and biopower, linking the Chinese to the West. For this reason, studying panda fandom through Western theories is not only necessary but also essential, as panda fandom belongs to the emerging life and experience of postcolonial scenarios within the evolving global capitalist regime.

This book is not a comprehensive investigation of giant panda fandom and its history in China. Rather, my specific perspective on panda fandom addresses what has been overlooked in Western scholars' studies of giant panda history – local myths, documentaries, personal narratives, and children's stories – providing the particular historical materialities that I engage with. While I examine how Chinese communities have developed their own responses to the West's exploitation of the giant panda in these historical documents, I also include interviews with panda fans at Panda Base to illustrate how panda fandom has reached its current state.

In April 2024, I delivered a seminar on early panda fandom in Finland, accompanied by a YouTube video showcasing the fervor and dedication of panda enthusiasts at Panda Base. The video, titled "August 28 Let the World See the Spirited Running of Chengdu's Youth for Hua Hua Watching", documented a daily vlog live-streaming young people racing for a precious 3-minute viewing opportunity with Hua Hua. Lasting 2 minutes and 55 seconds, the footage captured the vlogger sprinting alongside approximately 100 others through Panda Base. The giant panda Hua Hua, chomping bamboo atop an enrichment structure, appeared at the end of the video. Demonstrating adeptness with the ritual, the vlogger skillfully navigated past fellow runners while maintaining stable camera footage. Today, participating in the Hua Hua race has become a hallmark experience for panda fans who arrive early at Panda Base before its opening hours.

Following the seminar, a colleague expressed interest in joining the panda fan run, finding it entertaining. Another colleague remarked on the event's incredibility, contrasting it with European norms and forwarding a link to a European zoo's cautionary notice against running, particularly for children. I recalled the mornings when I encountered enthusiastic fans dashing along their routes. While Panda Base had implemented entry and enclosure fences to regulate visitor flow, there were no explicit regulations against fans running. Instead, the morning run had seemingly become a daily ritual and source of enjoyment, integral to Hua Hua's charm, supported and accommodated by Panda Base over time. Upon further reflection, it became evident that chasing after pandas was a common occurrence not only at Hua Hua's enclosure, as the crowd frequently surged with excitement, eagerly racing to witness an active panda.

Despite the fact that panda fandom, as I have suggested, is a result of Western–Chinese interactions over the past two centuries, this example demonstrates that the political, cultural, and economic uniqueness of the Chinese community also provides a distinctive ecology and materiality to panda fandom. In this context, panda fandom becomes a space where the Chinese community has negotiated its experiences and identities with Western powers over time. Chapter 1 explores the early Chinese–Western interactions from the discovery of the giant panda in the 19th and early 20th centuries. While the Chinese community, due to the influence of civil war, Western, and Japanese invasions during this period, was perceived as passive and disengaged from the West's scientific and trade interests in the giant panda, the Chinese community's involvement in the early establishment of the giant panda industry was significant. The Chinese learned to play specific roles in capitalist trade and cultural exchange. This forced engagement with foreign social and economic systems, although initially alien to the local Chinese community, sowed the seeds for the development of the panda fandom.

Chapter 2 examines the commercialization of the giant panda as a means of production within the framework of zoo attractions. It investigates the transformation of the Chinese community's relationship with the giant panda from passive involvement to enthusiastic participation. This chapter argues that the giant panda's appeal is not intrinsic to the animal itself but is a result of its integration into the evolving capitalist zoo attraction industry. In its early stages, this transformation relied on the exploitation of both the giant panda and the labor of the Chinese community. Using a children's story by the Chinese writer Chiang Yee, this chapter demonstrates how the giant panda's transformation into a zoo attraction in Western zoos necessitated the Chinese people's adaptation to and response to the pressures of capitalist industry and Western imperialism.

Chapter 3 explores cuteness as the central aesthetic attraction of the giant panda, considering how this aesthetic category reflects the animal's role as a means of production within the capitalist system. Cuteness, as a form of biopower, is more than just charm: it is, as illustrated by the WWF's giant panda logo, a cost-saving strategy. Cuteness links human and animal populations, which can be mobilized, managed, and regulated to perform specific tasks within the capitalist regime. This chapter demonstrates how the exploitation of both human and animal labor is aestheticized through the image of the cute. It shows how cuteness, in its alignment with biopower, serves as a tool of control.

In Chapter 4, I explore panda diplomacy as an extension of the giant panda industry, in which the Chinese nation began to play a more prominent role in the global industrial project. This chapter questions the Chinese community's involvement in the giant panda industry through the lens of panda diplomacy, highlighting how China's participation has significantly promoted the development of the panda industry. I reconsider the Chinese nation and political authority's role in panda diplomacy, showing that Western imperialist and colonialist ideologies often fail to recognize the Chinese nation's active engagement in this process.

The historical interactions between China and the West are crucial to understanding the emergence of panda fandom in contemporary China, which Chapter 5 addresses by examining the interconnected movements of capitalism, Western–Chinese relations, and biopower. This chapter argues that the panda, as a symbol, plays a pivotal role in the complex social and economic networks that sustain panda fandom. Drawing on Foucault's concept of biopower, I argue that the power structures underlying panda fandom enable the giant panda to become both a celebrity and a national symbol, but only by using the panda as a super-exploited means of production. Chapter 5 contends that panda–fan intimacy should not be understood as a natural attachment to the animal, but as a relationship shaped by contemporary social and productive relations.

Chapter 6 shifts the focus to my fieldwork on panda fandom, marking the transition from theoretical and historical discussions to contemporary observations. This chapter summarizes the data collected from the field, offering a bird's-eye view of the panda fandom. The following three chapters incorporate raw materials from this fieldwork to deepen our understanding of panda fandom. While previous chapters have laid out theoretical foundations, these chapters provide detailed, personal stories from panda fans who are actively engaged in the fandom today.

Chapter 7 explores the intersection of Chinese identity and giant panda fandom, tracing how the love for pandas has become intertwined with national identity in the context of Western–Chinese interactions. The phrase "only China has pandas" encapsulates how the image of the giant panda represents both national pride and the exercise of biopower. Chapter 7 demonstrates how Chinese fans, while expressing an idealized and unconditional devotion to both pandas and their nation, are also subject to the historical dynamics of Chinese–Western relations. Panda fandom provides a social sphere where these complex interactions are reified and understood in terms of Chinese identity.

Chapter 8 delves into the role of cuteness in panda fandom, exploring how it functions as a form of biopower. This chapter examines how cuteness shapes the daily lives of panda fans and aligns with capitalism's need for a disciplined and well-educated workforce, as well as cost-effective production methods. Fieldwork with panda fans reveals that the aesthetic of cuteness, as embodied by pandas, circulates capitalist relations through fans' identities, shaping social norms that aestheticize life. Cuteness, as a dominant aesthetic category rooted in biopower, perpetuates the economy and its cycles within the panda fandom.

Chapter 9 explores whether panda fandom can provide insights into building more-than-human worlds, despite the panda–fan relationship being embedded within capitalist, biopower, and national frameworks. This chapter argues that precisely because biopower is so powerful in shaping panda fandom, it also offers potential avenues for resistance and rethinking the nature of the panda–fan relationship. I show that panda fans have explored the possibility of a less exploitative and more encompassing relationship with the giant panda, characterized by a strategy of allowing the pandas to "just be there" and by reciprocal participation. Chapter 9 suggests that such a loving, nonviolent relationship with the giant panda may offer a model for more harmonious interactions with nonhuman beings.

During one of her visits to my Chengdu apartment in 2021, my mom mentioned that she wanted to see Hua Hua, a panda she had learned about online. She said, "I have grown very fond of this bear", referring to Hua Hua as one of the most beautiful and cutest pandas at Panda Base. I agreed with her – after all, all pandas are beautiful. But I had to turn down the task. "What? I didn't know we had this bear", I responded. I suggested she visit Panda Base herself to find Hua Hua. After all, with over 200 pandas at the Base, I wasn't involved in their daily care, so how could I know who Hua Hua was? My mom set out on her own search for Hua Hua, and when we spoke at dinner later, she excitedly told me how unique and beautiful the panda was. "How did you find Hua Hua?" I asked. She smiled and recounted how she had received help from security guards and fellow tourists to locate Hua Hua's enclosure. It turned out that panda fans knew the pandas, the enclosures, and the displays better than I did.

This was something I came to respect as I spent more time learning about panda fans. This book is not a survey study of panda fans or their pandas, but rather, I see myself as a scholar of panda fandom, imagining and exploring the more-than-human possibilities within it. When I first began this journey, I had no idea where it would lead. Initially, I hoped that studying panda fans might spark my own interest and devotion to giant pandas. However, that didn't happen, even as I neared the completion of this book. But as I ran alongside panda fans to conduct interviews while they rushed toward the queue for Hua Hua, I started calling myself "a fan of panda fans". Thus, this book emerges not only from my academic study but also from a unique experience of fandom—a fandom that, despite my distance from the pandas themselves, I have come to share in the passion and devotion that panda fans have for these remarkable creatures. Through this journey, I discovered that panda fandom is not only about the love for the pandas but also about how that love shapes interactions, identities, and experiences – creating a vibrant, evolving culture of its own.

1

GIANT PANDA

The Chinese history

Panda fandom is not solely about love and devotion for an animal. Instead, this chapter argues that affection for the panda has been shaped at the expense of the animal and her indigenous community, both human and nonhuman. This historical process went beyond a shared passion across nations and cultures; it was made possible because the giant panda, along with Chinese locals, became what Agamben and Sacer (1995) terms "bare life" – a form of existence that Western imperialist practices such as colonialism, science, and capitalism have continued to map and exploit since the 19th century. Panda fandom emerged as a response to this politicized form of life or, in Michel Foucault's terms, as a manifestation of "biopolitics". The enjoyment and appeal that fuel panda fandom are rooted in a calculated understanding of the panda's political existence, which was gradually disciplined and made suitable for display in international zoos and museums. The power to manipulate this politicized life form generated experiences of control and spectacle, becoming a productive force that facilitated capital accumulation and transformation. The "pandamania" and fascination that early audiences experienced in zoos and museums were thus symptoms of this orchestrated liveliness of the giant panda, intertwined with the history and people of China.

One key technique enabling the politicization of the giant panda has been the shift from an "axis of possession" to an "axis of spectacle" (McClintock, 2013, p. 231). The visual pleasure of witnessing the giant panda spectacle draws on preexisting patterns of fascination embedded in individual perception and social formations (Mulvey, 2013). For early panda fans, this socially constructed framework reinforced dichotomies such as human–animal, West–Chinese, industrial–wild, masculine–feminine, white–non-white, and scientific–traditional, all of which structured the visual experience of pleasure. This chapter explores the panda spectacle along four dimensions: scientific pursuit, commercial exchange,

DOI: 10.4324/9781003511229-2

narrative construction, and emotional reception. These aspects illuminate four interwoven dimensions of panda fandom and its historical context: the animal and ethnic culture as spectacle, the valorization of displacement in economic trade, the anthropomorphized creation of a new subjectivity through narrative, and the interconnected relationships of viewing, visitation, and consumption.

Science: Making the spectacle

One of the primary tasks for early Chinese biologists and scientists involved in giant panda research was to affirm the animal's place within Chinese history and tradition. Jinchu Hu, a leading Chinese biologist who collaborated with George Schaller to establish the Wolong Giant Panda Nature Reserve in 1978, contributed to this effort by publishing two scientific papers: One in 1981 to clarify the giant panda's historical record and another in 2009 to explore the animal's traditional Chinese names. Several scholars (何芬奇 & 孙前; 孙前 et al., 2008; 林鸿荣 & 林林, 2005), committed to linking the giant panda with China's cultural heritage, supported Hu's assertion that the giant panda "has always been a precious animal native to China since ancient times" (7) and is celebrated as a national treasure. This drive to connect an animal, whose scientific study only began in 1836, with the nation's traditional culture stems, interestingly, from the fact that references to the giant panda are sparse in China's historical documents. American historian Elena Songster (2018) describes the giant panda as largely a "modern phenomenon" in China, where scientific understanding, research, and widespread interest in the animal are relatively recent developments.

Perhaps one of the most detailed historical descriptions of a creature vaguely resembling the giant panda appears in *Liuba Ting Zhi*, a text by Qing dynasty literati He and Jiang (1842), who wrote the following about an animal they termed a "multi-color bear":

> The multi-color bear cave is fifty miles north of the shrine. It faces west. The bear living in the case is a creature as large as a cow, with a horse's head, a cow's tail, a dog's body, and four legs capable of walking like a human. Its back is black, its sides are white, and its neck and limbs are a mixture of black and white. It does not eat grains but consumes bamboo stems. It lacks five internal organs, having only one intestine, which is considerably large at both ends, resembling a belt that can be used to tie around the waist.

In this brief description, the beast was compared to several domesticated animals. Its distinctive white and black patches were noted, along with details of its diet. The literati also considered its utility to human society – although it was of limited practical use, its long intestine could serve as a waist belt.

However, the Chinese literati's account of the giant panda can be seen as part of a "cultural history" (Harper, 2013), which differs significantly from the scientific study and discovery of the animal. In his famous scientific account at Muping, French priest Armand David documented his encounter with the rare animal in his journal in different words (Aleteia, 2023):

> When I was on my way back to the church, I was invited to have a rest in a Mr Li's home. In this home, I saw the Panda's skin. It's big and beautiful, colored black and white. The skin was quite peculiar. Li told me that I would see this animal very soon, for his hunters were going to hunt this animal . . . it seemed that a new species in the science domain will be found.

Ten days later, David saw the body of a giant panda from the local hunters, which he acquired at a cost (Science Museums of China, 2023):

> They sold the young whitebear to me at a very high price. The body of the whitebear was all white except that the legs, the ears and the places around its two eyes are black. It has the same skin color as a grown-up bear that I have seen before. I believe it to be a new species, not only because of its skin color, but also because of the hair beneath its feet and other characteristics.

David further offered technical details and discussions about the animal:

> The young bear is entirely white, except for his four limbs, his ears, and the area around his eyes which are a deep black. Thus, we have here a new species of Ursidae that is very remarkable not only because of its color but also because of the hairiness under its paws.

Soon David shipped the panda's skin to the Natural Museum of Paris, addressing the giant panda in his letter to Alphonse Milne-Edwards, his contact at the museum (Nicholls, 2014):

> I have not seen this species in the museums of Europe and it is easily the most pretty I have come across; perhaps it will turn out to be new to science!

In the 1870s, Henri Milne-Edwards created the first formal scientific illustration of the giant panda, which was subsequently published. Within a decade, scientists and researchers in the West had documented the giant panda as a new species that had largely escaped Chinese cultural recognition, despite its 5,000-year history. In this scientific illustration, the giant panda gazes directly at the viewer, highlighting its unique features and contrasting sharply with the comparisons to domestic animals previously drawn by Chinese literati. Here, the panda emerged as a distinct subject, not reliant on relative descriptions but represented through exact measurements and data. Unlike the neutral observations of He and Jiang, David was captivated by the panda's beauty, calling it one of the most remarkable animals he had ever encountered. For David, the giant panda held significance not for its utility but for its unique existence and appearance, which resonated with his scientific inquiry.

David's encounter with the giant panda can also be contrasted with local religious perspectives, particularly Tibetan Buddhism, which shaped views on the natural environment and the relationship between local Tibetans and animals, potentially contributing to the limited accessibility of the giant panda to outsiders.

The giant panda inhabits several mountain ranges in south-central China, in the provinces of Sichuan, Shaanxi, and Gansu. These regions were historically governed by local leaders before the establishment of the Nanjing government of the Republic of China in 1927. The *Tusi* system, translated as "headmen" or "chieftains", was a governance structure implemented by central Chinese dynasties since the Yuan dynasty (1279–1368) to manage areas with diverse populations (Wang, 2019). This system was grounded in the principle of empowering "local people to govern local affairs" (Jia). This flexible approach allowed central authorities to maintain a cooperative relationship with local communities by granting titles and privileges to the Tusi, who then exercised leadership over their communities while acknowledging their place within the broader political framework. In return, the Tusi had obligations such as paying regular tribute and performing political, economic, and military duties. When Armand David decided to go to Muping and serve as a priest at Dengchigou Church, his visit was first recognized by the local Tusi (Hu, 1994).

The Tusi system maintained a delicate balance between the central authority and ethnic minority communities. Regions inhabited by ethnic minorities were often considered a lower priority for the central government, which sometimes fostered a cultural superiority over groups viewed as "barbarians beyond the border" (Wang, 2019). The central government frequently showed limited interest in developing these "bitterly cold frontiers" (Wang, 2019), instead focusing on rituals and maintaining face rather than directly exploiting border resources. Ethnic minorities were expected to show respect for Chinese mainstream culture, symbolically reinforcing the empire's status. As long as this cultural and political deference was upheld, the central government preferred to allow these regions a degree of autonomy in self-governance (Wang, 2019).

The Western Sichuan region, where the giant panda thrives, was self-governed under the Tusi system. This area served as a crucial link between the Chengdu Plain and Tibet, and Tibetan Buddhism held significant influence over the local community. In Tibet, the theocratic structure concentrated both spiritual and secular power, while in Western Sichuan, secular authority was vested in the Tusi system, with the monastic class reliant on secular support (Wang, 2019). In some cases, Tusi families themselves held spiritual authority. Nevertheless, Tibetan Buddhism reached local people through its teachings and dedicated followers.

Tibetan Buddhism, like other forms of Buddhism, presents an empathetic view of animals and other nonhuman beings. Animals are regarded as sentient beings capable of emotions and some degree of thought. Thus, they can experience suffering, and humans are encouraged to consider animals' needs when making ethical decisions. Barstow (2019) suggests that, although animals in Tibetan Buddhism are viewed as less intelligent than humans, the religion supports the idea that animals have experiences of the world similar to human experiences. Consequently, Tibetans and related ethnic communities consistently accord animals a significant moral standing. Hou and Jiang (1991) observed that Tibetan Buddhism in Western Sichuan fosters a spiritual bond between residents and the natural environment. For example, mountains hold profound significance, embodying dual meanings as both pathways to the heavens, full of mystery, and revered places where deities and ancestral spirits reside. In these mountainous regions, nearly every peak is tied to a captivating legend, and each is believed to be a manifestation of a deity guiding the lives of those nearby. During major events, such as warfare, intertribal conflicts, or religious ceremonies, men from surrounding villages traditionally gathered at sacred mountains for prayer rituals.

Terrone (2014) describes how Tibetan Buddhism regards the earth as a treasure, emphasizing an interdependent exchange between humans and their environment. In this "ecology of revelation"(Terrone, 2014, p. 460), any treasure removed from its place must be replaced with a "substitute treasure" to honor the local protective deities and sustain the environment. Zhang and Chen (2007) add that the Buddhist belief in reincarnation reinforces the principle of non-killing. In this view of cyclical rebirth, individuals are encouraged to cultivate empathy, as life moves through various realms; harming an animal or plant in this life may lead to experiencing similar suffering in future lives. Thus, Tibetan Buddhism provides a strong moral foundation for respecting animal life, discouraging harm to wildlife and other beings.

Studies have documented a close relationship between religious influence and wildlife conservation in natural habitats (Bhagwat et al., 2005; Cui et al., 2022; Landim et al., 2023; Li et al., 2014; Nijhawan & Mihu, 2020; Pachuau & van Schendel, 2022). For example, using avian community composition as a case study, Gesang et al. (2024) found that Tibetan Buddhist monasteries across the Qinghai–Tibet Plateau have become important sites for human–wildlife coexistence. The researchers concluded that the presence of these monasteries positively influences avian abundance, though not species diversity, compared to areas without monasteries. Similarly, Li et al. (2014) demonstrated that Tibetan Buddhist monasteries play a significant role in snow leopard conservation. The 336 monasteries studied in the Sanjiangyuan region provided protection for snow leopard habitat (8,342 km^2) through community norms and active patrolling efforts. The researchers observed that local herders refrained from hunting wildlife, with 42% citing Buddhism's view of killing as sinful. These scholars suggest that local religious beliefs and community devotion could be vital components in the design of conservation policies. When Armand David arrived in Baoxing, the local Tusi authorities had recently reiterated a decree banning hunting to support religious values and extend local governance.

Scientists' investigations into the historical records of the giant panda have largely focused on official documents produced by Han cultural and political authorities. In contrast, Tibetan Buddhism and the local cultures of the Tibet–Sichuan region – areas closely associated with giant panda habitats – have received comparatively less scholarly attention. Only a limited number of researchers have reported the rich traditions and cultural context of local Tibetan communities, with most studies relying on oral histories and informal accounts. For example, in his paper *Reinvestigating the Name of the Giant Panda*, Bo (2003) proposed that "Panda" is not a Western term but rather a phonetic translation from the local Tibetan dialect. Bo reached this conclusion based on correspondence with one of his readers. However, Lin and Lin (2005) later dismissed this theory, suggesting instead that the panda bear was likely known as "Měng" in the local dialect before David's encounter with the animal. For the locals, the adoption of the name "panda" rather than "Měng" was shaped by ongoing interactions with Western hunters and scientists in the region (林鸿荣, 2005). Over time, with annual hunting activities, the name "panda" became widely recognized, to the extent that government officials and locals alike took pride in using it.

It appears that tracing the existence or cultural significance of the giant panda may be more effective through the oral histories and legends of local communities, although these sources can be challenging to access and verify. However, examining Tibetan Buddhism and the influence of Han cultural dominance provides some insights into the historical obscurity of the giant panda in Chinese records. First, in Tibetan Buddhism, the giant panda was not perceived as an isolated entity separate from its ecosystem as David Armand conceived in his scientific journal. Instead, Tibetan beliefs viewed the giant panda, along with other animals, as an inseparable part of the natural world. The panda and the mountains it inhabited were regarded as deities, sacred to local culture. Moreover, the mountains' profound religious significance to the locals limited interactions between people – hunters included – and the giant panda. Respect for the sin of killing animals and a ritual requirement for replacement further protected pandas and their habitats. Thus, even though Ernest Wilson (1913) noted that local hunters had extensive knowledge of the land during early Western expeditions in China, the giant panda continued to be a revered and respected presence within the mountains.

In his seminal work, *The spell of the sensuous*, Abram (2012, p. 63) closely examined various indigenous cultures worldwide, noting that, unlike Western civilization, these cultures have often developed a "remarkable solidarity with the lands that they inhabit, as well as a basic respect, or even reverence, for the other species that inhabit those lands". Tibetan Buddhism, it seems, is one such culture that has fostered a relatively balanced and sustainable relationship with the giant panda and the surrounding mountains. Abram emphasizes that these indigenous cultures have maintained a mutual relationship with the land through oral traditions and narratives. Similarly, the bond between the giant panda and its habitat may be seen as part of a rich, oral exchange between local Tibetans and their environment – a relationship best understood through careful study of local languages, rituals, traditions, and art forms.

Additionally, the historical central position of Han Chinese culture and the influence of Han cultural dominance over minority groups (Harrell, 1993; Tam, 2023) – as exemplified by the introduction of the Tusi system in Tibetan regions – created challenges for communication between the Indigenous Tibetan community and Han authorities. Because Han authorities often idealized a collective identity that valorized Han Chinese civilization, the oral traditions and cultures of Tibetan communities were sometimes overlooked. The Tusi system, with its self-governance model, ultimately reinforced the cultural divide between these cultures, prioritizing political control over cultural integration.

Consequently, neither Tibetan Buddhism nor Han Chinese culture offered formal means and texts to document the existence of the giant panda. In Tibetan Buddhism, the giant panda was not an isolated figure but rather part of a land that the local community engaged with respectfully and drew sustenance from. Within Han Chinese culture, the giant panda was compared to domestic animals or humans, yet the animal's image remained unclear due to the political dynamics between Han and Tibetan communities. Armand David's encounter with the giant panda, however, was not constrained by either Tibetan Buddhist or Han Chinese perspectives. Unlike Tibetan Buddhism, which saw ecological integrity as central to the panda and its habitat, David viewed the animal as distinct from its environment – the panda was a unique specimen, largely separate from the mountains it inhabited. Rather than conceiving the giant panda as integral to its natural surroundings, David's first encounter involved the animal's displacement, through a panda skin he observed at the home of a local man named Li.

David's approach to the giant panda also sharply contrasted with the perspective of Han Chinese authorities. The French priest actively engaged with lands, communities, and places that Han authorities had traditionally regarded as remote, uncivilized, or peripheral. The Dengchigou Church where David resided was one of the first Western missionary stations established in Sichuan, intended to influence local customs; what initially drew David to Dengchigou was the region's rich biodiversity (Nicholls, 2012). As Schiebinger and Swan (2007, p. 2) note, science and Western colonialism often collaborated, providing "access to ever farther-flung regions of the globe". Once the giant panda became a scientific subject and bare life – open to study and observation by science and Western culture – the quiet and secluded life of Dengchigou and its surrounding communities began to change. The panda soon came to embody a network of social relations linking China and the West over the next two centuries.

Capitalization: Trade and blood

During my first weeks working at Panda Base, I visited a biologist involved in the giant panda reintroduction program to ask for books on the animal's history. Our brief conversation about the panda's past ended abruptly when he said, "Outsiders know nothing about the giant panda's story. Much of what's written is misleading. Did you know the first live panda smuggled abroad was labeled as 'a dog' to pass through customs?" This story – of a national treasure of China registered as a dog during inspection – struck me deeply. As someone who had spent half of her formal education in China, I could strongly empathize with the nationalist sentiment and sense of humiliation in this tale. Yet, what unsettled me more was the implication that I, with my additional Western education, was somehow disqualified from fully experiencing this historical pain. At that moment, I was an "outsider", someone who knew little of the animal's and country's "true" history. It felt as if China's traditions were just beyond my reach.

Yet, the story of Su-Lin, the panda labeled as a dog in 1936, was not a "skeleton in the cupboard" that my colleague hesitated to reveal. Instead, it felt like a call, nudging me toward a deeper understanding of the giant panda and reconnecting with my "Chineseness". At that moment, I felt my own lack of knowledge about this part of "Chinese history" keenly, wondering how a single story of a panda from 1936 could make such a powerful impact on someone educated and well-traveled in her 30s.

Further research shows that scholars widely agree that China's recognition of the giant panda and its rise to international fame was a delayed response to experiences with Western imperialism and colonialism. In *The Way of the Panda*, Nicholls (2012) divides the panda's history into three phases, the first of which spans from David's encounter with the giant panda in 1869 to the establishment of the People's Republic of China in 1949. Nicholls describes this period as one where the nation "adopts a passive role in the panda story . . . trying to free itself from the grip of Western colonialism, to replace its downtrodden imperial establishment with a new republic and to fight off unbelievable Japanese aggression". In *Panda Nation*, Songster (2018) echoes Nicholls, suggesting that the giant panda – despite existing in China for millions of years – is largely a "post-1949 phenomenon", as "the general populace did not take a strong interest in it until after 1949".

Both scholars imply that China was either too preoccupied with domestic crises or too indifferent to recognize the giant panda's symbolic importance early on. Following this interpretation, my colleague's story about the panda and my reaction to it could be understood as an expression of nationalist or patriotic sentiment, similar to how many non-Western and Indigenous groups respond to Western imperial dominance. If the giant panda was, indeed, shaped by the rise of the People's Republic of China, then perhaps this 1936 tale of humiliation could be dismissed by my colleague and me as a historical footnote. Yet the story seemed to cast a spell on me, one that might affect anyone who identifies as Chinese. What is this underlying connection that binds my colleague and me through the story of the panda? If the panda-dog story still resonates so powerfully with us, why would Nicholls and Songster suggest that the Chinese experience of this history was purely passive or indifferent?

Exploring this gap between the Chinese experience and Western scholarly perspectives goes beyond the current scope of this book. Readers might find more thought-provoking answers in the work of cultural critics like Chow (2002, 2003) and Shih (2001). Here, I return to the central figure of this book – the giant panda – and ask how engaging with this nonhuman participant might illuminate some of the questions surrounding anthropocentric interpretations of West–East interactions. An undeniable fact from this initial phase of the panda's rise to global fame is its transformation into a global spectacle. While studies and documentaries on the giant panda (Croke, 2009; Nicholls, 2012; Songster, 2018) detail Western explorers' expeditions to obtain pandas – leading to their sensational appearances in Western zoos and museums – this transformation was not achieved without crucial collaboration from the Chinese.

German historian Jürgen Osterhammel (1986), aligning with Chinese economist Yanan Wang, introduced the concept of *China dependency theory*, illustrating how early Chinese interactions with Western powers included local collaboration that facilitated Western political and economic penetration. Viewed in this light, the commodification of the giant panda and its transformation into a universal symbol were the results of a cooperative dynamic between China and the West. Far from being passive in this capitalist transformation, the Chinese played a central role, often amid conflict and hardship, contributing, as Chow (2002, p. 34) puts it, through "the ethnicization of labor", to the panda's eventual valorization on the global stage. To delve deeper into this process, Donna Haraway's (2013) concepts of *lively capital* and *encounter value* provide a useful framework.

Lively capital and encounter value

In her interpretation of Marx's classical theory of capitalism, Haraway (2013) argues that humanist teleology – centered on human labor power and the self-making of humanity – remains the focus of Marx's work. Haraway suggests that when nonhuman beings participate in the commodification process, they introduce the concept of *encounter value* into Marx's framework of use and exchange value, even though nonhuman entities remain oppressed, unrecognized, and uncredited within anthropocentric capitalist relationships. Thus, Haraway emphasizes the importance of understanding animal–human relationships within the historically specific context of *lively capital*.

Reflecting Haraway's rhetorical use of categories such as prisoners, the disabled, and strays to describe oppressed animal members, scholars widely agree that animals in capitalist exchange often endure greater suffering than human laborers due to their inability to transcend the capitalist system and its alienating practices (Painter, 2016; Torres, 2007). Within this line of research, animals' roles in capitalist production are compared to those of human workers. Similar to the working class, animals are subjected to a set of power relations reinforced by capitalist hegemony, yet they derive little benefit from their participation. Painter (2016) suggests that animals are even more alienated within capitalist relations than human workers, as they are entirely deprived of freedom and life experiences, existing solely to satisfy human commercial and personal interests. As Painter (2016, p. 334) states:

> since this process denies animals the possibility of satisfying even their most basic interests, such as eating food that is natural to their diets, being able to defecate in an area separate from where they sleep and eat, being able to socialize, or being able to move freely about, let alone live a flourishing life, from a non-speciesist perspective it is reasonable to argue that the alienated existence of animals is worse than it is for human wage workers, given that many (though not all) of the latter, while certainly unable to flourish within capitalism, are at least able to enjoy positive, meaningful experiences outside of their work-life.

In his further examination of Haraway's notion of lively capital and encounter value, Barua (2016, p. 4) suggests that rather than measuring animal participation in capitalist production through a human-centric concept of labor, human–animal encounters should be understood as a "process of value generation where the bodies, ethologies, and liveliness of an animal make a difference to, and are constitutive of, those very relations that render or mobilize it as a commodity". Traditionally, human beings are seen as the sole agents enabling productive conditions, relegating animals to the role of raw materials in the value-creation process. This anthropocentric view of labor and production, Barua argues, alienates animals from the capitalist circle. Instead of reducing animal participation to an inferior category or mere raw materials, Barua (2016) proposes that animals engaged in capitalist accumulation represent a heterogeneous form of labor that has been devalued, fetishized, and objectified under the dominance of human labor.

Barua's (2016) approach to acknowledging nonhuman participants in generating lively commodities involves a two-step perspective. First, while animal participation differs from human labor, it remains integral to the value produced; animals and their bodies are thus "inert" components in lively capital (Barua, 2016, p. 14). Second, human labor provides conditions that foster lively capital's growth rather than directly "making" value (Barua, 2016, p. 14). By considering animal labor as a constitutive component within lively capital and positioning human labor as a "conditioning" rather than a "making" force, Barua's approach allows animals to occupy more than a mere raw material status in the production process.

In addition to recognizing animals as integral to producing lively commodities, Barua (2016) argues that animal charisma offers a second commodifying factor, forming "an organizing force based on cult and followings, an order of enchantment that is more alluring than the mundane and bureaucratic workings of the economy" (Barua, 2016, p. 15). Collard (2020, p. 18), in her study of the global exotic pet trade, describes this phenomenon as "double fetishization": the first layer involves conventional commodity fetishism, while the second reflects romanticization and anthropocentric affection for exotic animals. Collard's concept of exotic animal fetishism parallels Barua's notion of animal charisma, reinforcing the animal's labor and participation within the production process.

In his conceptualization of animal exploitation in capitalism, Stache (2020) concurs with Barua (2016) that animals are not equivalent to human laborers. However, rather than framing animal involvement in capitalism through ideas tied to labor, Stache (2020) states that animals are regarded as natural resources and raw materials within the production process. Thus, animals are not treated as labor forces but are integrated into capitalism's social relationships of production. Stache (2020) posits that, much like nature and the working class, animals play a distinct role in the capitalist system, where the relationship between capital and animals differs from that between capital and labor. As with nature, which serves the capitalist venture, animals are best understood as a "means of production privately owned by capital" (Stache, 2020, p. 414).

Viewing animals as the means of production and private property of the capitalist, Stache (2020, p. 412) argues that animals are "super-exploited" within capitalism.

> First, they (animals) are completely appropriated by capitalists from their first day on earth and either they work or get worked on, for example, in vivisection, to their death or they die through the hand of a slaughterer. Second, the status quo of animals with regard to the bourgeois social relations guarantees capitalists that there are no discounts of profits in favour of animals.

Unlike the working class, who receive wages in exchange for labor, animals are not compensated in the capitalist market, which allows capitalists to maintain profits without any deductions for animal welfare. As private property, animals are denied the political freedoms available to human proletarians. Through this private ownership and role as a means of production, Stache (2020) argues, animals remain external to the capital relation, while capitalists gain full economic benefit from this externality. By distinguishing animals from the working class, Stache (2020) suggests that animals are not commodities but products in capitalism; only the application of waged labor transforms an animal product into a commodity. This positions animals in capitalism similarly to slaves, yet, as Stache (2020) points out, animals lack the capacity for collective resistance and conscious rebellion, unlike enslaved humans. Consequently, animals are incapable of transcending their conditions within capitalism, remaining as a means of production and products that exist outside the capital relation.

Stache's (2020) analysis of human–animal relations in capitalism complements Barua and Haraway's conceptualizations of lively capital, providing a critical lens on the intersections of classism and speciesism. As Haraway (2013) notes, animal–human encounters are intertwined with the power dynamics of gender, nation, and ethnicity. In exploring how the giant panda transitions from a product to a commodity of attraction shaped by the labor of both Chinese and Western actors, a dual examination of colonial and imperial agendas – agendas that also exploited Chinese laborers within lively capital – becomes essential. This labor, both physical and emotional, has largely gone unrecognized, even within academic discourse. For this reason, I now turn to Rey Chow's notion of the "ethnicization of labor" to examine how the commodification of the giant panda has been propelled by the contributions and efforts of the Chinese community.

The ethnicization of labor

Beginning with Louis Althusser's renowned text "Ideology and Ideological State Apparatus", Chow (2003) argues that "ideology", as a condition of production, aligns itself with the capitalist process of production. Chow (2003, p. 22) explains that Althusser is not only referring to "quantifiable value (wages) but also a 'historically variable minimum'—'needs'", meaning the "mental attitudes, wishes, sufferings, and fantasies of individuals involved in the process of active production". According to Chow, ideology shapes a form of consumption and reception that enables people to buy into and indulge in cultural offerings. Thus, the acceptance of an ideology or culture, no matter how "passive" and "natural" this reception may seem, always contains a "responsive, performative aspect" (Chow, 2003, p. 22). Feminist scholars like De Lauretis (1984) and Silverman (1983) have used the concepts of "suture" and "interpellation" to describe how female subjects respond to oppressive conditions by constructing ideologies and illusions as a survival mechanism.

Chow (2003) further applies this materialist approach to modern Chinese literature, particularly in its portrayals of women. She observes that popular literary schools of the modern age, such as "Mandarin Duck and Butterfly", succeeded in the market because of their adept engagement with capitalist terms. In this reading, Chow (2003) argues that Chinese "tradition" and "Chineseness" became accessible cultural signifiers that the public could identify with in modern China. This display of tradition allows the capitalist market to reproduce the existing relations of production while nurturing the forces of production in return. Chow (2003, pp. 82–83) concludes that the butterfly stories served as a kind of commodity that enabled a "powerful and successful line of exchange with urban readers through their emotional and epistemic crisis".

In *The Protestant Ethnic and the Spirit of Capitalism*, Chow (2002) further explores ethnicity's entanglement with capitalist relations, clarifying the connection between forces of production and the prevailing relations of production. She uses the term "the ethnicization of labor" to explain how people marked as ethnic others are relegated to socially inferior and low-level labor positions. As these marginalized workers become aware of their commodification due to the exchange of their labor, they often attempt to resist this transformation through protest. This resistance enables the working class to transcend their immediate social class.

However, drawing on Max Weber's insights into the integration of Western secular religion and the capitalist regime, Chow (2002, p. 47) points out that resistance and protest, "when understood historically, are part and parcel of the structure of capitalism; they are the reasons capitalism flourishes". She argues that capitalism has fabricated a politics of ethnicity such that non-Western communities remain in captivity. Attempts by ethnic others to transcend these boundaries – through protest or aspirations for emancipation – further the visibility, currency, and circulation of ethnic identities on a global scale. Chow concludes that ethnicity becomes a symptom of the entwined relations of capitalism and its biopolitics.

Chow's discussion on ethnicity reveals that identifying with one's ethnic heritage involves engaging with the social relations that uphold the capitalist production system. This engagement is inevitable under the omnipresence of capitalism and is essential for survival among ethnic communities. However, this identification also marks the subject's commodification. While the ethnic subject may wish to resist this transformation, this resistance ultimately upholds the capitalist structure, as it becomes an economically beneficial form of productive labor within capitalist accumulation. Consequently, the capitalist system creates a "material-cum-spiritual" reward cycle from which ethnic subjects cannot escape.

Haraway's notion of "lively capital" and Chow's concept of ethnicization both aim to highlight the labor and value of marginalized groups within capitalist systems of commodification. Both scholars examine how these systems depend on productive forces and the social relations that commodify nonhuman animals and ethnic communities. Yet, while Chow (2002) draws a parallel between the captivity of ethnic subjects in capitalism and animals in modern zoos, the subjective experience of commodification differs between human and nonhuman participants. This prompts an exploration of a local giant panda myth, as recounted by Dake Hu (1994) in "The Debut of the Giant Panda in Baoxing", to illustrate how the ethnicization of local Chinese communities and the animalization of the giant panda were simultaneous capitalist projects to transform the giant panda into an international spectacle.

The debut of the giant panda in Baoxing

Dake Hu's chapter appears in *Giant Panda in Baoxing*, a collection edited by the History and Culture Committee of the Political Consultative Association of Baoxing County. On the cover, it specifies that the collection is for "internal use", and its preface is written by Keming Qi, Baoxing County's Party Secretary in the early 1990s. Hu (1994) describes himself as a writer dedicated to researching giant panda history since 1971 and notes his involvement in a local TV show on giant pandas. The chapter is, according to Hu, the result of his 20-year documentation and careful verification. It is divided into four sections: an introductory overview of giant pandas and their history, followed by narratives on three Western figures pivotal to the panda's debut – Armand David, the Roosevelt brothers, and Ruth Harkness. The section on the Roosevelt brothers is particularly relevant here.

Despite Hu's suggestion that the compilation had been carefully selected and verified, his narrative of the Roosevelt brothers is best understood as a local myth, due to the discrepancies between Hu's story and the historical records of the event. For example, Hu claims that the Roosevelt brothers visited Baoxing several times over the course of 3 years to hunt the giant panda, but in reality, their stay lasted less than a year. Additionally, Hu's story includes the exhibition and death of the first giant panda the brothers hunted, a claim that lacks verification in any credible documents. This blending of history and fantasy suggests that Hu's account is more of a myth, retold by the locals as the giant panda became a known animal to them. According to Chow (2003), it is precisely this fantastical structure and fabrication of the story that give the myth its value. As the local community struggled to make sense of their first encounters with Westerners and the capitalist agenda, these interactions gave rise to local myths that, in Chow's (2003) words, "suture" the crises in belief and practice. Myths and ideologies are created and recreated amid historical contingencies through purposeful acts of memory, intention, and labor (Slotkin, 2024). Thus, the fabrication and revision of myths serve as a coping mechanism to navigate these crises.

Hu's story of the Roosevelts exemplifies how the local Chinese people struggled to bridge the gap between the colonial and capitalist transformation of the giant panda and the collapse of feudal Chinese culture. In this light, Hu's story stands as perhaps the only culturally available example for understanding how the Chinese engaged with the giant pandas and ultimately contributed to their commodification. Below, I provide a summary of Hu's version of the Roosevelt brothers' story.

In March 1926, the Roosevelt brothers, accompanied by three assistants, a translator, and two Muslim cooks from Chengdu, arrived in Shanggan Valley, Longdong Town, Baoxing County. Staying with Haitang Wang, a local landlord, the Roosevelts gifted him an American rifle, two Chinese guns, and 100 silver yuan. Although they largely avoided interaction with locals, they engaged a few young men to help carry equipment and scout hunting paths, paying each two copper coins daily.

The Roosevelts and their team first traveled to Lujin Valley. All the men were required to stay with the Roosevelts, regardless of the weather conditions. During the day, the hunting team explored the mountains and shot any animals they encountered. After 5 days of searching, the team finally came upon a giant panda. Fearing the panda might escape, Theodore shot it in the hind leg. The team then carried the giant panda to Wang's house, and four strong young men spent the next 3 days transporting the panda to the Catholic church in Ya'an. The locals later learned that the brothers had shipped the giant panda from Leshan back to Chicago. However, the panda died the day before its scheduled exhibition in Chicago. Theodore was deeply distressed by the animal's death and had the giant panda preserved as a specimen, which was later displayed at the Natural History Museum in Chicago.

Disappointed by the panda's death, the Roosevelts returned to Wang's family in 1927. Wang suggested they search Baixiong Valley, where they captured several birds. In the bamboo forest of Baixiong Valley, they took a prized bead, which locals interpreted as foreigners "stealing treasures".

In 1928, the Roosevelts told Wang that a one-horned monster was living in Wutan Mountain and that they wanted to help the locals kill it. The brothers asked Wang for more local hunters to assist them. Although Wang had never seen the monster himself, he believed in its threat and decided to support the Roosevelts. More than ten local young men once again joined the Roosevelts in hunting the creature. However, when the team arrived at Wutan Mountain, they found it peaceful, with no sign of the monster. The Roosevelts then instructed the team to search for the giant panda instead. With the locals' help, the Roosevelts quickly had a successful hunt, killing 14 golden snub-nosed monkeys in no time. At the site, the Roosevelts skillfully dissected the monkeys, performing intricate techniques that the local hunters were unfamiliar with.

The Roosevelts continued their pursuit and finally trapped a giant panda by luring her with the scent of a sacrificed sheep. The brothers caged the panda and began carrying it down the mountain. It was only at this moment that Wang realized the Roosevelts' true intention was to capture a giant panda. The hunt for the one-horned monster had been nothing more than a deceptive ruse. Furious, Wang gathered the villagers, led them to surround the Roosevelts, and confiscated their prey. He and his team then forced the Roosevelts out of Shanggan Gou. As news of the incident spread, the valley where the foreigners were expelled became known as "Gan Yang Gou" (Chasing Foreigners Valley). Even today, the people of Baoxing still use this name for the place.

An initial reading of Hu's story suggests that the exchanges between the local Chinese and the Western hunters were primarily based on monetary terms. For instance, early in the Roosevelts' interactions with the landlord Wang and local hunters, their relationship was maintained through the exchange of payment and gifts. This exchange of payment and labor continued into the final stage, when the Roosevelts successfully captured the giant panda. It seems that this exchange facilitated the locals' trading of animals with the Western hunters. Wang and the local hunters accepted the payment and gifts as compensation for killing the animals. To a larger extent, it could be argued that the local Chinese closely collaborated with the Roosevelts in hunting and killing animals.

As Osterhammel (1986) pointed out, in its semicolonial and semifeudal state, China's ruling class – including big landlords, compradors, and bureaucrats – while not entirely identical to foreign powers, based their political influence and economic growth on imperialism. Wang's collaboration with the Roosevelts was made possible by the landlord's proximity to Western imperialist figures and his oppression of other local hunters. The Roosevelts paid wages to the local hunters through Wang, who remained the central figure mediating and benefiting from this Chinese–Western interaction.

Before delving into the transformation of the giant panda into a nonhuman commodity, it is essential to understand how the Chinese locals themselves were also commodified through their engagement in the capitalist relationship. It is through this double commodification – where both the locals and the giant panda contributed their labor in diverse ways – that the giant panda became one of the most attractive animal spectacles in history. This analysis of the commodification of the locals is based on Chow's discussion.

Capitalist production relies on the replication of its productive forces and the existing relations of production. These two aspects warrant separate examination. In terms of productive forces, landlord Wang and the locals contributed their knowledge of the land, as well as their skills in load-bearing and hunting. The identifiable tasks that the Chinese performed included translation, cooking, carrying loads, hunting, guiding, and providing accommodation. In contrast, the Roosevelts played the roles of managers, guests, hunters, scientists, and entrepreneurs. This division of labor creates a clear distinction between the locals and the Roosevelts: While the locals performed the more rudimentary tasks involved in hunting the giant panda and were largely unaware of the final product of their labor, the Roosevelts possessed the tools of commodification, such as guns, dissection techniques, specimen preparation skills, and an entrepreneurial mindset geared toward monetary exchange. For both the Roosevelts and other Western scholars of the giant panda's history, the labor of the locals remained largely uncredited, even though they were integral to the process.

As Chow (2002, p. 34) observes, the concept of ethnicity is intimately tied to hierarchical divisions of labor. The "otherness" produced within privileged societies contributes "labor toward the accumulation of capital yet at the same time receives the least of its rewards". In Hu's account of the giant panda, the active participation of the locals – without whom the hunting of the giant panda would have been nearly impossible – is depicted as "bitter labor" (苦力 in Chinese), which supported the more advanced, skilled labor of the Roosevelts. Here, the capitalist division of labor reinforces a hierarchical structure between the Westerners and the Chinese, with the Chinese occupying the lower, "othered" position.

The existing social relationship that further consolidates the structure of the productive forces is the readily available ethnic category of the Chinese. Hu's story stands as a testament to the locals' identification with their ethnic tradition, which provided what was culturally available to them during the crisis of capitalist commodification. To understand this labor process of becoming "the Chinese", it is necessary to turn to the climax of the story – the Roosevelts' deception of the locals.

Wang's rejection of the Roosevelts' deceptive claim was the pivotal moment that deteriorated the previously established relationship between the two parties. In the story, the Roosevelts were the deceivers, as their claim to be hunting a local monster ultimately turned out to be a ruse to capture the giant panda. However, as mentioned above, the Roosevelts and Wang had initially collaborated closely through the exchange of money, gifts, and local resources. The Roosevelts' true purpose – hunting the giant panda – became apparent to Wang and the locals on the very first day of their encounter. This 2-year collaboration between Wang and the Roosevelts could have conveyed the brothers' intentions to Wang. This suggests that, despite the Roosevelts being the initiators of the deception, Wang and the locals must have shown a willingness to be deceived. It is this surrender to the Roosevelts' obvious deception that led Wang to organize a local hunting team to assist the brothers. In this light, Wang and the other hunters had, at this stage, knowingly decided to be deceived.

This decision to be deceived marks the moment when Wang and the locals' class consciousness emerged – their awareness that their labor and contributions had been alienated and were being incorporated into the colonial and imperial project of commodifying the giant panda. Wang's leadership and the locals' assistance were crucial to capturing the giant panda, but as locals cohabiting with the animal, they had no control over the commodification process. As the giant pandas became specimens in Western museums and exhibits in zoos, the locals had no access to, nor relationship with, these products. The Roosevelts' deception thus provided a moment of painful realization for Wang and the locals: They themselves, like the giant pandas, had been transformed into commodities.

This awakening motivated Wang and the locals to drive the Roosevelts out of Shanggan Gou, where the brothers had been hunting the giant panda. On the surface, the Chinese had won a battle against the foreign invaders and defended their homeland. However, as Chow (2002) argues, it is precisely this desire to rebel and protest from within one's social class that completes the circle of capitalist production. The moment the Roosevelts were labeled "foreign devils" and "outsiders", the locals simultaneously identified with the Chinese identity. In this coupled experience of alienation through productive labor and identification with Chinese culture, the locals became integrated into the capitalist production circle.

From ethnicization to animalization

Juxtaposed with the commodification and ethnicization of the Chinese was the inclusion of the giant panda into the capitalist system. This double capitalization process, while aimed at creating value for the West, marginalized both the ethnic and nonhuman participants in this process. While I will later turn to examine the production of the giant panda as a commodity, a further question remains: What was the relationship between the simultaneously commodified Chinese and the giant panda? Did the Chinese, inescapable from the capitalist mechanism, apply the same exploitative relationship to the giant panda? Or, in this mutually marginalized otherness under the regime of capitalism, did the Chinese construct a different set of social relationships with the animal?

In Hu's story, the giant panda is most vividly captured in three scenes. The first scene involves the death of the first giant panda in Chicago before her exhibition night and the capture of the second giant panda, which was based on a story of deception. While the second capture contains some historical truth, matching the Roosevelts' trophy hunting, the first capture and the transportation of the animal to Chicago seem to be the most imaginative aspects of the narrative. On the one hand, the exportation of the giant panda to Chicago contradicts the recognized history in which Ruth Harkness brought the first live giant panda out of China. On the other hand, the death of the Chicago giant panda on the night before her exhibition debut suggests that the animal becomes an embodiment of sentiment and emotions unique to the Chinese community. Drawing on the framework of lively capital, I will examine the inclusion of the panda into the productive force and the existing social relationships that facilitated the animal's transformation into a commodified product.

In terms of labor forces, both Barua (2016) and Stache (2020) suggest that animals in capitalist relationships perform tasks, albeit in ways that differ from human labor. While Barua (2016) maintains that animals perform heterogeneous labor, Stache (2020) shows that nonhuman beings differ from human laborers in two key ways. First, animals do not receive wages as compensation in the capitalist relationship. Instead, they are freely exploited as a means of production and raw materials, with capitalists benefiting significantly from this exclusion of animals from the wage-based economy. In Barua's (2016) terms, the giant panda's tasks described in Hu's story included her liveliness, bodily (non)existence, and ethological behaviors. However, unlike Wang and the local hunters, who received meager compensation for their labor, the giant panda received no such remedy for her captivity.

Second, the giant pandas, shot by the guns of the Roosevelts, lost their mobility and freedom the moment they were captured. This forced immobility and removal from their natural habitat illustrate that the giant pandas were not laborers who could freely exchange their labor for compensation. Rather, the giant panda's loss of freedom was a form of suffering that the locals did not experience. Unlike the humans, who participated in the capitalist system through their labor and received at least some form of compensation, the giant panda's suffering was a direct result of the exploitation of her body without any acknowledgment or remedy.

What further sets the giant panda apart from the local Chinese is the existing social relations that support the reproduction of the productive forces. In the last section, I showed that the Chinese community was transformed into laborers through their physical contributions and identification with "Chineseness". In contrast, the giant panda did not engage in a wage–labor exchange with the Roosevelts. Rather, the giant panda was considered a free resource, one that the brothers could possess and incorporate into the production process at no cost. According to Brenkert (1979, p. 123), capitalist private property is the "power possessed by private individuals in the means of production which allows them to dispose as they will of the workers' labor power". Once claimed as the Roosevelts' private property, the giant panda granted the brothers power over the Chinese locals, whose labor had helped realize the capitalist design.

In this sense, the commodification of both the locals and the giant panda reveals the intricate dynamics of exploitation and alienation in capitalist systems. The Chinese locals became the working class offering labor to exchange wage, and the giant panda was possessed by the Western capitalist as private property, the means of production. While the locals' labor was instrumental in capturing the panda, they remained alienated from the end product, much like the giant panda itself. The commodification of the animal, however, was not simply only a matter of extracting labor but also of establishing dominance over both human and nonhuman subjects, consolidating power within the Western imperialist framework. Through this double commodification, both the Chinese and the giant panda became entangled in the capitalist machine, but each occupied distinct and unequal positions within it.

The possession of private property provides capitalists with the means to exert power over the working class and their labor. As Brenkert (1979) points out, for capitalists, private property acts as an antithesis to labor, as it is meant to increase productivity for greater profit. However, the workers must sell their labor power to capitalists in exchange for wages. In this capitalist framework, the local Chinese as laborers and the giant panda as a means of production are linked through a paradoxical relationship of both dependency and exploitation. On the one hand, the Roosevelts' project on the giant panda transformed the animal into private property that the Chinese did not own or control. On the other hand, the local labor was vital to their survival during a period of internal conflict, providing them with much-needed compensation in exchange for their contributions.

In local Tibetan Buddhism, the giant panda was considered a minor deity, a sacred being inhabiting the mountains. The panda was an integral part of the natural surroundings, revered as an entity within the ecosystem. However, the Roosevelts' hunting expedition created an entirely different relationship between the locals and the animal. This shift was profound and complete, to the extent that the first giant panda captured was destined to meet her death on her debut day. By "fated", I mean two things. First, for the giant panda, her death as a super-exploited means of production in the hands of the capitalists became her inevitable fate. As the story unfolds, even in death, she was not liberated from the capitalist system. Dissected and turned into a specimen, the panda continued to provide service for the profit of the capitalists, perpetuating her role within the capitalist framework. Her death on the night of her debut was portrayed as a last act of agency, an expression of freedom in the face of her exploitation. It seemed that this was the only form of resistance the giant panda could exert against the oppressive and encompassing capitalist system. Yet, her death, much like the protest of the local Chinese community, only confirmed her integration into the capitalist system.

It is important to recognize that the "fated" death of the giant panda is a fictional element in Hu's narrative. In other words, this fantasy dedicated to the panda should be viewed as the Chinese locals' attempt to come to terms with their transformed relationship with the animal. This anthropomorphized retelling of the panda's death reflects a deep emotional response to the new social dynamics and power structures that the Chinese locals found themselves enmeshed in. The death of the panda, imbued with symbolic meaning, is a reflection of the local community's historic condition – their transformation into commodities at the disposal of the West. The dramatic emotional expression through the panda's death serves as a historic marker of the exploitation the locals experienced, linking their fate to that of the animal they had once revered.

While an interpretation grounded in shared comradeship between the exploited (the locals and the panda) might suggest some form of empathetic reciprocity, this view overlooks the paradoxical relationship between the local Chinese as labors and the panda as a means of production. Rather than establishing a bond of mutual empathy, recognizing the relationship between the Chinese community and the giant panda could be seen as naturalizing the panda's super-exploitation. The narrative of the panda's death, therefore, becomes more than a story of shared suffering; it represents the Chinese local community's historicity – their role as the working class, transformed into commodities under the capitalist system.

The giant panda's death is emblematic of the working class's protest. Just as Wang and the local hunters expelled the Roosevelts from the valley, asserting some form of resistance to foreign exploitation, the panda's death reflects the struggle of the working class to subvert the capitalist class. In this way, the panda's fate mirrors the working class's desire for emancipation, as the animal's death marks the final, irreversible integration of both the Chinese laborers and the giant panda into the capitalist system. The tragic death of the panda, then, is not only a loss for the animal but also a reflection of the broader narrative of exploitation and the commodification of both human and nonhuman labor under capitalism. It is through this lens that the giant panda's death is seen not merely as an individual tragedy but as a symbol of the historic transformation of both the local community and the animal into commodities within the capitalist framework.

Contextualized within the capitalist relationship, the deception story co-created by the Roosevelts and the Chinese locals offers important insights into the dynamics surrounding the capture of the second giant panda. The deception, as perceived by Wang and the other locals, was framed as the facade of hunting a mysterious local monster, a narrative that concealed the true objective of capturing the giant panda. This facade of the monster was not only a cover for the Roosevelts' real purpose but also reflected a deeper process of alienation – the transformation of the Chinese locals into laborers whose contribution was detached from its products. This deception, then, serves as a critical lens through which to examine the changing relationship between the Chinese and the giant panda.

Before the hunt, Wang and the locals deemed it fair and a "good deed" to hunt the monster due to the harm it had caused. In this light, the hunt for the local monster, based on the moral grounds of public welfare, represented a return to what was familiar to Wang and the locals. However, the story struggles to provide the same justification for hunting the giant panda. For instance, after the hunt, Kermit Roosevelt observed that the hired local guides refused to bring the animal into their village, assuming that "the giant panda is a sort of minor deity among them" (Heller, 2015). The Roosevelts (Roosevelt & Roosevelt, 1929, p. 263) also noted their Indian guides' reaction to the hunt: "The giant panda, from all that we could learn, is not a savage animal. After the shooting, our Kashmiri shikaries remarked that he was a Sahib, a gentleman, for when hit he had remained silent, not calling out as a bear would". Clearly, the Roosevelts easily dismissed these local objections to killing the giant panda, as the pursuit of profit and imperialist goals had justified the hunt.

It is through this inability to come to terms with the giant panda that the animal's status as a "means of production" continues to dominate the local Chinese. Songster (2018) observes that "even after the giant panda attracted interest outside China, local attitudes toward the giant panda indicate that people who lived in its habitat had little use for it". For American historians, this remark is a testament to the giant panda's role in the story of the rise of the People's Republic of China. Yet, precisely because the giant panda, at this stage, was an exclusively and privately possessed property of the West and the scientific community, it was completely out of the locals' control. Their alienation from the means of production ensured that, as laborers, they "had little use" for the animal. While Hu's story brings the complex relationship between the local Chinese and the giant panda back into focus, it is notable that this dynamic is not addressed fully, where the paradoxical intersection of the means of production and the laborers remains unexamined.

Conclusion

Starting with the unsettling history of the giant panda, retold from the perspective of the Chinese, provides a framework for understanding the emergence of the giant panda fandom in both Western and Chinese cultures. In this chapter, I draw on science and capitalism as two major forces that have altered the Chinese relationship with the giant panda. The scientific encounter between David and the animal created an image of the giant panda that can be explored through accuracy and precision. This scientific spectacle presents a different giant panda from the one seen in Chinese Han culture and local Tibetan Buddhism. I propose that the giant panda's obscurity in Chinese history is tied to the Chinese authority's political and cultural strategies toward the remote locals and ethnic minorities in history.

Following the Western construction of knowledge about the giant panda, starting with its scientific image, capitalism provides a thorough framework for transforming the relationship between the local Chinese and the animal. Despite Western scholars suggesting that the Chinese were passive and indifferent to the giant panda before 1949, I argue that this period marks the critical stage in which the giant panda became relevant to capitalism. This transformation relied on the locals being redefined as ethnic others and commodities through their labor, and the super-exploitation of the giant panda as a natural resource freely available to the West. The Western capitalists, such as the Roosevelts, possessed the giant panda as private property and wielded power over the labor of the local Chinese through their ownership.

I demonstrate that despite both being marginalized and othered in the capitalist relationship, the connection between the Chinese and the panda was not one of companionship or empathy. Rather, trapped within their historicity, the Chinese were unable to transcend their class consciousness and struggle with capitalist oppression. As a means of production, the giant panda formed a paradoxical relationship with the local Chinese. In this capitalist framework, the giant panda exerts power over the Chinese, as she can dispose of their labor force, while the Chinese become dependent on the animal, exchanging their labor for survival.

Staging the giant panda as a means of production, both benefiting the capitalists and being worked on by them, distinguishes the giant panda from the commodified Chinese in this early history. While the giant panda does not contribute labor and is then treated as a spectacle, it produces products (Stache, 2020) that are further appropriated by the capitalists and the market. Previous studies have focused on how scientists (Nicholls, 2012) and conservationists (Schaller, 1994; Songster, 2018) have labored diligently with the giant panda and its products to create commodities. The next chapter will shift focus to the working process that transformed the giant panda into a unique commodity in the form of visitor attractions. The first step toward giant panda fandom began with the animal's transformation into an attraction.

2

MAKING AN ANIMAL ATTRACTION

> "At the time, the panda hunter was an international sensation, and the panda himself once drew more than 53,000 visitors when first displayed at the Brookfield—a single-day tally the zoo has never again matched."

In 1936, the American designer Ruth Harkness acquired the giant panda Su-Lin in Baoxing and transported the animal to the United States. She later sold Su-Lin to Brookfield Zoo for $8,750, using the proceeds to fund a second expedition to China in search of another panda, Mei-Mei. By the time Su-Lin arrived at Brookfield Zoo, the public was primed for her arrival, and she received a sensational welcome. The zoo declared the acquisition of the giant panda as "the most important single achievement in collecting animals in modern times" (Croke, 2009). Ruth Harkness became "the darling of the media" (Croke, 2009), while Su-Lin herself drew over 300,000 visitors to Brookfield Zoo within just 6 months of her debut (Heller, 2015). Notable figures such as Helen Keller and Al Capone were among the crowds who came to see Su-Lin (Heller, 2015).

While the giant panda's prominence in scientific research and conservation has been thoroughly studied (Nicholls, 2012; Songster, 2018), her role as a zoo attraction and her rapid establishment as a global sensation in Western zoos have received comparatively little attention. To Western audiences, the giant panda's charisma, rarity, and charm seemed to make her an inherently captivating figure. In the case of Su-Lin, the panda appeared to be the missing piece in the zoo experience, seamlessly completing the spectacle of attraction and drawing enthusiastic Western crowds.

In contrast, what seemed like a natural appeal of the giant panda to Western audiences and scholars was not immediately evident to the Chinese locals. Scholars such as Nicholls and Songster have observed that the Chinese were initially indifferent to

DOI: 10.4324/9781003511229-3

the giant panda's charm and attraction, as the country was preoccupied with urgent, life-or-death conflicts, including war and political upheaval. However, as recounted by Croke (2009), Ruth Harkness's effort to transport the giant panda Su-Lin out of China in 1936 began to stir anti-Western sentiments, fueled by an emerging sense of Chinese national pride. Harkness faced significant obstacles, including detours between Chengdu and Shanghai, and extensive negotiations with high-ranking Chinese authorities and Chinese Maritime Customs to secure Su-Lin's departure for the United States. This dynamic highlights the disparity in how the panda's appeal was perceived by the Chinese and the West. While the panda's beauty, charisma, and cuteness captured the Western imagination early on, these qualities were a belated recognition for the Chinese. Instead, the panda's symbolic importance for the Chinese was sparked by a "rising sense of indignation" (Croke, 2009), as they witnessed their country's resources being taken by Western powers. This indignation marked the beginning of a deeper connection between the Chinese and the giant panda.

This chapter proposes that the transformation of the giant panda into an attraction should be critically examined by questioning the presumed "naturalness" of its appeal. This presumed naturalness underpinned the enthusiastic reception of the panda by Western audiences in 1937 and fueled subsequent phenomena like *pandamonium* (Chinoy, 1975) and *pandamania* (Scott, 1987). The giant panda's appeal has often been taken for granted as an inherent and physical fact. Yet, if the panda's charm was truly an inherent characteristic, why did it fail to evoke a similar sensation among the Chinese during those early years? Rejecting an Orientalist perspective, what factors might have rendered this "natural" attraction ineffective outside the Western context?

The appeal of the giant panda, I argue, can be more effectively understood through Marx's (2000) insights into capitalism. Marx (2000, p. 184) suggests that capitalism advances by leveraging "natural predisposition (e.g., physical strength), needs, accidents, etc.". Within this framework, the panda's perceived attractiveness operates as one of these "natural predispositions" that capitalism redefines and naturalizes. The panda's physical charm, reinterpreted as an inherent quality, resonated deeply with Western audiences already immersed in capitalist systems. Simultaneously, the rising national sentiment in China – opposing the West's relocation and capitalization of the panda – was similarly reframed as incidental "accidents" or localized "needs" within the broader capitalist narrative. This dual process highlights how both the panda's charm and the Chinese reaction were appropriated within the logic of capitalist exploitation. To further develop this materialist interpretation of the giant panda's appeal, it is necessary to revisit the cues outlined in the previous chapter.

Product and commodity

The previous chapter demonstrated that the giant panda and her products functioned as a means of production owned by Western explorers from her initial emergence in Baoxing. Since her debut at Brookfield Zoo, the giant panda Su-Lin has remained a major attraction. Before delving further into the concept of attraction and its role within the capitalist market, it is crucial to distinguish between *product* and *commodity*. These distinctions are essential because the means of production and commodities are integrated into capitalism in fundamentally different ways. Understanding these differences helps challenge the often-assumed notion that the giant panda, as an attraction, is merely a commodity. Contextualizing the giant panda within the commodification process, as I will argue, bypasses her status as a super-exploited entity in capitalism.

In Marx's framework (2000), a commodity in the capitalist market is defined as a good or service produced specifically for exchange. A commodity is characterized by its dual nature: *Use value* and *exchange value*. Use value refers to the utility or usefulness of a commodity in satisfying human needs or desires, often tied to its physical properties and functions. Exchange value, on the other hand, represents the relative value of a commodity in relation to others. It is an abstract and quantitative measure derived from the average labor time required to produce the commodity. As commodities are produced for profit, they are intended to be freely exchanged in the market.

The *means of production*, particularly natural resources, differ significantly from commodities. According to Marx (2000, p. 249), the means of production represent the "subjection of Nature's forces to man". Entering the capitalist form, the means of production are perceived as capital, existing as entities external to human existence. Marx contends that the means of production are alienated from humanity because they embody "the particular type of labor, labor which is leveled down, fragmented, and therefore unfree" (Marx, 2000, p. 95). This alienation highlights the means of production's externality to capitalist relations, allowing them to be owned privately and utilized as property. Such externality enables the means of production to direct labor processes and forces with varied characteristics.

Marx identifies commodities and the means of production as two foundational elements of capitalism. Their key distinction lies in their function: Commodities are produced for exchange, while the means of production are privately owned and separate from the market. The means of production are tools for organizing labor processes and forces that generate commodities for market exchange.

Reflecting on these distinctions, Stache (2020, p. 412) asserts that "non-human animals as such are not commodities". Instead, animals, their parts, or the products of their labor are incorporated into the use value of commodities. A critical step in transforming an animal or its products into a commodity lies in the application of wage labor. Human labor adds exchange value, introducing an abstract dimension to the animal or product that allows it to circulate freely in the market. Stache (2020, p. 412) emphasizes that "prioritising the commodity form to conceptualise animal is not accurate since it obscures the relation which makes animal and their parts a part of the use value of a commodity". In this sense, animals can contribute to commodities but are transformed into commodities only through waged human labor.

Following Marx's definition and Stache's note, given the giant panda's inability to circulate freely in the market, it is problematic to assume that the panda alone is a commodity exchanged between capitalists and consumers. Instead, her status as a super-exploited entity, in the form of an attraction, suggests that the giant panda should be considered part of the attraction, while the attraction itself is the commodity circulating in the market. This leads to a critical question: What, then, constitutes the human labor force that transforms the giant panda into an attraction? Answering this question is essential to understanding her role within capitalist production.

The attraction as a social relation

The main difficulty in answering this question is the dynamic role human labor occupies in the creation and operation of these attractions. This dynamic requires a nuanced approach to avoid oversimplification or distortion. To explore this, I turn to Dean MacCannell's seminal work *The Tourist*, one of the most frequently cited texts for its insights into the concept of tourist attractions. My use of MacCannell's work is not an attempt to extend its theoretical scope, as many researchers have already provided rich expansions (Cohen, 1988; Knudsen & Waade, 2010; Moore et al., 2021; Rickly, 2022; Rickly-Boyd, 2012; Wang, 1999). Instead, I focus on its continued relevance to the tourism industry, specifically in terms of its "recurrent theme in accounts of tourist experiences of place and culture and as a quality of tourists' experiences" (Moore et al., 2021, p. 1).

This enduring relevance suggests that MacCannell's work is not only a theoretical abstraction but also an integral part of how tourist attractions are experienced. The text itself is deeply embedded within the structure of the attraction, shaping the social relations that make the very concept of a tourist attraction possible. From this perspective, MacCannell's work continues to resonate with scholars, tourists, and the tourism industry because it aligns with the material reality of how attractions function, both socially and experientially.

In his study, MacCannell (1976) introduces the concept of authenticity in the sightseeing industry, suggesting that authenticity has been a desire and need of tourists, who consistently search for its existence in attractions. The attraction attracts, in other words, because it contains a promise of authenticity – the core need of modern beings in general. However, MacCannell further shows that authenticity can always escape the grasp of tourists because of the inherent dualistic structure of the front stage and back stage in the attraction. In tourist attractions, the front stage is carefully managed and designed, aiming at curating the tourists' experiences and meeting their expectations. In contrast, the back stage refers to what remains inaccessible to tourists but is the everyday life and work of the locals/hosts/Indigenous. The attraction marks a promise of authenticity that can never be realized within the attraction and its own system. To solve this irreconcilable contradiction within the attraction, MacCannell suggests that its commodity form, a symbolic representation of everyday life in modern society, has to be transcended through an "immense accumulation of reflexive experiences" (MacCannell, 1976, p. 23).

Central to MacCannell's understanding of attraction is the reflective transcendence that modern individuals must perform as a remedy to the inherent lack of the commodity form of attraction. The appeal of authenticity is the fetishistic appearance of attraction that operates in everyday life in modern society. In the front and back stages dictum, the modern individual's experience with work is curiously dichotomized. On the one hand, the process of work at the back stage becomes the display of work at the front stage and is hence available to be consumed by the individual. On the other hand, his experience with work is restricted to the front stage through the work display, demanding his transcendence. MacCannell (1976, p. xvi) concludes that the tourist, the sightseer who views the attraction, is an "early postmodern figure, alienated but seeking fulfillment in his own alienation".

Looking at MacCannell's tourist attraction through a Marxist lens mandates the question of what is considered the labor process, the means of creating value and commodities in the capitalist market. MacCannell (1976, p. 36) reflects that labor division fragments individuals from their social relations, and the solution is "to search for his identity or soul in off-the-job activities". Because attractions such as "programs, trips, courses, reports, articles, shows, conferences, parades, opinions, events, sights, spectacles, scenes and situations of modernity are not determined by the amount of labor required for their production", MacCannell (1976, p. 36) suggests that the "capacity to deliver an experience" creates value in attraction. MacCannell's attraction, in this light, becomes a subjective experience in the tourists' search for their own identity, and it further argues that this search, the promise it delivers, finally determines whether the individual's experience is valuable or not.

In MacCannell (1976, p. 23), work and work relations do not generate value, but their collaboration with the mechanism of "museumization", which transforms them into a "work display", allows tourists to make a sensible and valuable experience. MacCannell sees the attraction as a symbolic system, with work and work relations as the inputs and the experiential tourists as the final products. In his further investigation of the mechanism of the museum, MacCannell (1976, p. 84) clarifies in the following words what can be considered or used as the "work and work relations" in the work display:

> This ideal of the museum is one that contributes to the unification of the modern world, to control over tradition and over nature. Modern museums and parks are anti-historical and unnatural. They are not, of course, antihistorical and unnatural in the sense of their destroying the past or nature because, to the contrary, they preserve them, but as they preserve, they automatically separate modernity from its past and from nature and elevate it above them. Nature and the past are made a part of the present, not in the form of are unreflected inner spirit, a mysterious soul, but rather as revealed objects, as tourist attractions.

> The museums, monuments, parks and restorations of modern society indicate that the staging of otherness and the organization of disparate elements in collections and representations into a single design of modern making, with the modern world flowing past its designated attractions, renders history, nature and traditional societies only aspects of the structural differentiation of the modern world, and not privileged aspects either, or at least no more privileged than the other attractions.

What becomes manifest is that in MacCannell's tourist attraction, the tourists and their subjective experience of the attraction provide the yardstick against which the attraction's value is measured. The attraction attracts, in other words, because the tourists are actively searching for an identity beyond their own alienated working conditions. This, however, does not mean that the tourists' own working conditions and experiences have been transformed by the tourist attraction. Instead, we need to note a key difference between the work and work relations displayed by the attraction and the tourists' own work experiences. The work and work relations that can be displayed as an attraction, as the above quotation shows, are what need to be preserved – the history, tradition, nature, and the other. Instead, the tourists' own work experience, a relation that continues to alienate the tourists, can only be transcended instead of preserved at the cost of the displayed work.

With this close examination of MacCannell's attraction, we can now turn to the making of the giant panda attraction in the 1930s. The question we need to explore is why the giant panda instantly attracted millions of Western visitors while its charm was largely unrecognized by the Chinese. Studies on the giant panda and its history have documented in detail the Western nations' marketing campaigns for obtaining a giant panda from China (Croke, 2009) and China's preoccupation with internal war and Western colonialism (Nicholls, 2012), but, according to the theory of MacCannell, the market promotion and war are mere symptoms of the attraction rather than its causes.

Since the visitor experience with the attraction creates the only source of value, the public was attracted to the giant panda because the encounter with the animal holds a promise of transcendence. For Western visitors, the transcendence and authenticity, through encounter with the giant panda, build up an identity and a unique touch with the modern society. For example, Osgood (1931, p. 1) reported the Roosevelts' acquirement of the two giant pandas in these words:

> The Roosevelts' habit of success, however, did not fail them, and they have brought to the Field Museum not the only giant panda specimen in the world, as some accounts have stated, but the only complete and perfect one and the only one killed by white men.

The celebration of nation-state power and the establishment of ethnic hierarchy made the giant panda attractive to Western audiences. In contrast, as noted in the last chapter, the local Chinese were still struggling to come to terms with Roosevelt's deception while increasing indignation toward the West's invasion started to surface. The encounter with the giant panda experienced by the Western tourists fulfilled the promise of a national identity, ethnic superiority, and scientific success.

It follows that the giant panda, in MacCannell's attraction, is a "work display" of social relation that collectively shapes the nations, ethnicity, and scientific inquiries. At the center of the attraction, the giant panda curiously occupies a marginal place – her display can be a valuable commodity into which the giant panda herself and the Chinese counterparts are integrated. The museumization of the giant panda made the animal attractive by preserving her as evidence of nature, the Chinese tradition, a distant past, and an unattainable history. In transforming the giant panda into an otherness, the animal marked an attraction captured by Western museum and zoo displays.

While MacCannell (1976, p. 19) states early in his text that he has significantly departed from Marx's idea of labor and commodity since he believes that a commodity has a "systemic quality" that generates value through its exchange, his observation of the "work display" resonates with Marx's critique of commodity. MacCannell shows that the displayed work and work relations are those that demand preservation, and the preserved also means the work's otherness. Thus, only what can be considered nature, history, and tradition could be displayed, while they remain external to capitalist exchanges. MacCannell's observation that the displayed work and work relations have been the othered nature, history, and tradition has only affirmed Marx's notion of commodity and capitalism – the work and work relations displayed as attraction are a means of production rather than commodities themselves. As a means of production, the work and work relations remain the private property of the Western capitalists. They are also a means of organizing labor forces and hence the devoted search and identification of tourists.

This observation shows that a Marxist and materialist approach to tourist attraction is not only possible but necessary. Viewing the giant panda through MacCannell's idea of attraction, the animal's super-exploitation can be underpinned in the following two ways. On the one hand, centering the value creation process on tourists' individual experiences suggests that the attraction can only be referenced through a subjective and human perspective. This anthropocentric view of attraction has legitimately excluded the participation of the giant panda and her production from the attraction. On the other hand, the giant panda displayed is reduced to otherness, such as nature, tradition, and history, and is hence excluded from engaging in further social relations, such as exchange. In this conception of the panda's double exclusion – from the working class, whose labor receives compensation, and from commodity exchange on the capitalist market – MacCannell's attraction mirrors the capitalist relationship supporting its operation.

The attraction: In materialist terms

In Marx's lens, the attraction is a commodity exchanged for the purpose of gaining profit, with two key components: The capitalization of private property as a means of production and the working class's labor forces. The labor forces create value for the attraction, while the means of production are exploited through the private property system. For the giant panda attraction in zoos, the materialist framework of attraction boils down to two questions: 1) How does the giant panda serve as a means of production in the attraction? 2) How does labor, namely, the value creation process, organize around the means of production, the giant panda?

1. The giant panda as a means of production

After Ruth Harkness acquired the giant panda Su-Lin and exported the animal to the United States, she sold the animal to Brookfield Zoo for $8,750, a sum that funded her second adventure to China to acquire another giant panda (Croke, 2009). This exchange between Harkness and the Brookfield Zoo facilitated the giant panda's role as a temporal commodity. However, the giant panda's commodity form should not extend beyond the zoo's purchase stage: the imperial adventurers, on the one hand, and the zoo institutions, on the other, exchanged their private properties with the intent of further realizing a profit. Harkness's exchange with the Brookfield Zoo was a trade between two capitalists, and the exchange ensured the optimal exploitation of the means of production. Once possessed by the Brookfield Zoo, Su-Lin's confinement as private property remained unchanged. Su-Lin was not, by any means, a freely circulated commodity between the zoo and the tourists. Yet, once acquired by the zoo, the giant panda Su-Lin was used as a unique and precious means of production incorporated into the zoo's attraction system for the public.

Differentiating the giant panda as the means of production from her existence as a commodity in the tourist attraction is key to understanding the animal's super-exploitation in the sightseeing service. On the one hand, the animal experiences the loss of her individuality and freedom from the natural habitat as soon as she enters the capitalist system. This loss of integrity with her natural environment means that in capitalism, her existence is fragmented. For example, the giant panda specimen, the giant panda in an enclosure, and the giant panda on postcards are all capitalist products, each enfolding a piece of the animal into a specific commodity. To an extent, this unlimited and ongoing engagement with the giant panda represents how people interact with nature under capitalist conditions. Just as the giant panda can be produced and reproduced without boundary, nature, it seems, also provides resources and inputs without limit. For another, the giant panda was not elevated to the status of human laborers, whose participation in the production process is compensated. In contrast, studies (Nicholls, 2012; Schaller, 1994; Songster, 2018) show that in animal enclosures, zoos, scientific labs, and breeding centers, provisions for the pandas are kept to a minimum before the animal's welfare became an international concern. The one-way conversation between human participants and the giant panda ensures that the animal can contribute to profit generation as a freely accessible resource.

Viewing the giant panda as the means of production rather than a commodity in the attraction product allows us to answer the question at the beginning of this chapter: Why was the giant panda an attraction to Western audiences in the 1930s, while her natural appeal was clearly lost on the Chinese audience? It suffices to suggest that Western society, with its established capitalist conditions and structure, interweaves the animal seamlessly into a capitalist network. The Western audience was attracted to the animal as the panda attraction serves a ubiquitous role. To this extent, the attraction is a concentrated representation of the human participants' exploitation of natural resources and their alienated existence from nature.

In contrast, at a time when the Chinese were still under the influence of a semicolonialist and semifeudal society, the giant panda, as the last chapter suggested, represented less a direct human-exploiting-nature relationship but a juxtaposition of the "ethnicization of labor" (Chow, 2002) for the Chinese. This, as I will return to later in this chapter, does not mean that the Chinese would not develop an exploitative relationship with the giant panda, but that the Chinese had to resolve this complex of abjection (Kristeva, 2024) as a priority. A needed step toward this question is a better understanding of how the giant panda, as a means of production, was not only exploited but also organized labor, ultimately contributing to the valorization of the giant panda attraction.

2. The organized labor in attraction

Despite MacCannell (1976) attributing the value of attractions to the subjective experiences of tourists and the transcendence these attractions promise, a Marxist critique suggests that the individual and personal search for authenticity in a tourist attraction is only a commodity for exchange in its capitalist form as the attraction itself has its own organization and structure, one that has evolved within capitalist relations. For an animal attraction like the giant panda at the Brookfield Zoo, the panda serves as the means of production for the attraction, suggesting that further human labor is required to realize the exchange and abstract form of this commodity. On the surface, the claim of necessary human labor in tourist attractions seems to be at odds with the ideology of the leisure class (Veblen, 2007) and the nonworking activities (MacCannell, 1976; Urry & Larsen, 2011) typically associated with these attractions. However, only human labor enables the transformation of animals and their products into commodity. How can we approach the necessary labor process within the visitor attraction?

One of the most explored theoretical paths to understanding the attraction and its mechanisms is the framework of the display (MacCannell, 1976), exhibition (Macdonald, 2007), or museumization (Bennett, 2013). John Berger (2015, p. 26), in his classical book *About Looking*, succinctly describes the mechanism of the display:

> Zoos, realistic animal toys and the widespread commercial diffusion of animal imagery, all began as animals started to be withdrawn from daily life. One could suppose that such innovations were compensatory. Yet in reality the innovations themselves belonged to the same remorseless movement as was dispersing the animals. The zoos, with their theatrical decor for display, were in fact demonstrations of how animals had been rendered absolutely marginal. The realistic toys increased the demand for the new animal puppet: the urban pet. The reproduction of animals in images—as their biological reproduction in birth becomes a rarer and rarer sight—was competitively forced to make animals ever more exotic and remote.

Berger's approach to zoo animal attraction resonates with MacCannell's (1976) observation that attractions could be seen as the "display of work". For Berger, zoos are venues where not only the animals but also the work involved in making animals marginal are on display. In Berger's terms, animal attraction is not only about the animal but also about the social relations that have captured the animal. MacCannell's discussion also aligns with this, where the work involved in the display is the marginalization of what is presented – the history, the other, the tradition, and the past. This degradation of natural resources, such as the giant panda, and the absence of non-Western involvement have been key focuses in scholarship.

In his study of museum exhibitions, Macdonald (2007) summarizes the history of the encounter among science, technology, and exhibition, based on three phases. The initial phase focuses on the emergence of modern science museums, the rise of collecting during the Renaissance, and the development of scientific knowledge frameworks in the 17th century. Museums and exhibition techniques were primarily focused on generating a particular kind of taxonomic knowledge controlled by authoritative voices. Visitors to these museums were limited to figures of social, political, and intellectual importance, whose personal nobility added to the museum's establishment as a source of authoritative knowledge.

The subsequent phase examines the evolution of public science museums and the rise of world fairs, particularly during the late 18th and 19th centuries. The opening of museums to the public was influenced by themes such as the formation of nation-states, colonial expansion, the development of scientific and museological ways of seeing the world, and the application of these to various domains of life. Museum exhibitions not only became venues for showcasing narratives of modern progress but also served as technologies through which modernity – and its democratic ideals, social differences, exclusions, and contradictions – was mediated (Macdonald, 2007, p. 8).

This is the key stage when museum objects moved beyond private scientific displays to become public attractions. The rise of nation-states redefined populations as citizens, transforming museums into symbols of national identity and progress (Macdonald, 2007). Public visits to museums became a form of civic education, and the act of "exhibiting" extended beyond museums to include major world exhibitions like the Great International Exposition in London in 1851. During this period, specialized public museums in natural history, ethnography, science, and art flourished.

The emergence of modern ways of seeing the world through representation or world-as-exhibition (Mitchell, 1991) provided a primary means of "apprehending and colonizing reality" (Macdonald, 2007, p. 9). As Mitchell (1991, p. 93) notes:

> These means of coordination were something particular and physical, offering what Michel Foucault has called a microphysical power; a power that worked by reordering material space in exact dimensions and acquiring a continuous bodily hold upon its subjects . . . this power was something meta-physical. It worked by creating an appearance of order, an appearance of structure as some sort of separate, non-material realm.

In this orderly setting provided by museum exhibitions, the public entering the museum space is transformed into visitors. Mitchell (1991, p. 19) explains that the visitor is a curious "combination of detachment and close attentiveness", attracted to the representation of the world. Here, museums have evolved into a means of production that can shape both the physical and mental experiences of the public on a large scale.

The final segment addresses developments from the 1960s onward, highlighting shifts in display methods within science museums and the emergence of industrial heritage sites and science centers. This stage evolves around the dominant themes of exhibitions since the second phase and an increasing need to represent many different and minor voices (Macdonald, 2007). New technologies of display, innovative interpretative methods, and new concerns are incorporated into museum exhibitions to support and validate the inclusion of multicultural differences. Macdonald (2007) suggests that localized, cosmopolitan, and reflexive approaches are distinctive ways to contextualize science in museum exhibitions at this stage.

This schematic history of exhibitions in museums shows that since the late 18th century, exhibitions have become attractions that draw on the particular viewing experiences of visitors and the representation of the world. The attraction denotes a way of looking at the world and organizes bodies, populations, and objects based on the principles of the exhibition. For Mitchell and Macdonald, exhibitions ceased to be mere private collections for scientific curiosity and became a means of production, mediating and organizing the public's relationship with nation-states, political agendas, and scientific endeavors.

Similarly, in his project investigating the birth of the modern zoo, Rothfels (2002) defines the German zookeeper Carl Hagenbeck's innovations in zoo design and animal exhibits as the critical event that shifted the display of encaged zoo animals to roaming lives in naturalistic enclosures. This transformation, from the menagerie, where captive animals are collected for display and the aggrandizement of the owner, to zoological gardens, where scientific endeavors are privileged and public education is imperative, in 19th-century Europe and America, Rothfels (2002, p. 37) reflects, celebrated "the tasks of enlightened and bourgeois progress in the world" and showcased the "optimism, power, and ambitions of the new bourgeois elite". Notably, the progress into the zoological garden, where animals are shown inhabiting their natural environment, concurs with the distinctive way of looking at the world in museum exhibitions. While it is not hard to detect themes of state-nation citizenship, colonial ambition, and scientific categorization in zoological gardens, their coupling with the technique of representation and exhibition demonstrates that modern zoos, like their museum counterparts, have become a significant means of production in producing commodities such as animal attractions. The schematic history of museums and zoos shows how the two organizations evolved over time into an accumulated structure of means of production through consistent engagement with human labor and intelligence, natural resources, infrastructure, knowledge, and, fundamentally, the technology of exhibition.

Taking modern zoos and museums as organizations with specific physical structures, social relations, and facilities capitalizing on profit attractions as commodities suggests that zoos and museums are parallel to the industrial factory in Marx's (2000, p. 429) terms as "a form of social labour organized on machinery as its material basis". In this light, Carl Hagenbeck's innovative approach to zoo exhibitions was, to an extent, not too different from Henry Ford's introduction of assembly line production, for both aimed at advancing the existing resources and methods relevant to their final commodities. Ford achieved his goal through the introduction of the assembly line as an automatic system to reduce the cost of labor. Similarly, Hagenbeck conducted a meticulous study of zoo exhibitions and introduced naturalist and ethnographic exhibits to the public. As Rothfels (2002, p. 86) notes, Hagenbeck reflected that visitors attracted to the zoo display were largely drawn to "the absolute novelty of not only this presentation, but of such presentations whatsoever". To better understand how zoos or museums, as accumulated forms of social labor organized through the technology of exhibition, generate profit, it is essential to investigate attractions as commodities exchanged for profit.

The look as a bio-technique

The tourist attraction, the end product and commodity of institutions such as zoos and museums in capitalist societies, in Marx's terms, aims to gain profit through its exchange in the market. Like any commodity, an attraction is also characterized by both use value and exchange value (Marx, 2000). Moreover, an attraction belongs to commodities exchanged between people based on the recognition of the abstract form of social labor as the only source of value. The attraction conceals this abstract form of social labor time and is accessed by tourists through its fetishistic images. In the case of the giant panda attraction, the fetishism surrounding it creates the illusion that the giant panda is naturally attractive and that tourists visit zoos solely for the sake of the giant panda. However, to obtain exchange value, the attraction must first fulfill its use value. In other words, the attraction needs to satisfy specific human needs and desires in order to exploit its value for exchange. The following examines this unique need that tourists find in attractions by closely examining exhibitions and displays, the new technology that museums and zoos have incorporated since the modern era.

Following Macdonald's (2007) schematic analysis of the history of the museum, it can be said that the modern zoo has achieved its goals via a similar trajectory. Mitchell (2007: 7) citing Theodor Adorno, refers to the zoo as "a product of nineteenth-century colonial penetration of the Orient", which "paid symbolic tribute in the form of animals". Themes that modern museums (Macdonald, 2007) have focused on – the nation-state community, colonial and imperial expansion, and the museological ways of seeing the world – can find parallels in the birth of zoos. These themes are meant to meet certain desires or needs of the tourists, who are the consumers of the attraction. For example, the nation-state community provides citizenship, symbolizing national identity and progress. The colonial and imperial agendas showcase raw materials, territories, and nature requiring further order and governance. The museological perspective offers clues to the framing arrangements. While I will explore these themes in more detail based on the study of the giant panda attraction, I specifically focus on how the use of exhibitions for this mode of production has helped make the attraction a distinctive commodity needed by the modern public.

In *The Age of the World Picture*, Martin Heidegger (1977) explains how the picture, or exhibit, provides the primary means of engaging with the contemporary world. The world picture, in Heidegger's (1977, p. 81) words,

> does not mean a picture of the world but the world conceived and grasped as a picture. What is, in its entirety, is now taken in such a way that it first is in being and only is in being to the extent that it is set up by man, who represents and sets forth. Wherever we have the world picture, an essential decision takes place regarding what is, in its entirety. The Being of whatever is, is sought and found in the representedness of the latter.

The world-as-exhibition suggests that people's lives – their activities, relationships, and labor – are linked to the world through an exhibitive relationship. This dominance of the visual relationship with the world has been elevated to the state of Being, suggesting that entry into modern life depends on grasping the world through a picture. In Heidegger's terms, the exhibition, because of its dominance and prevalence, becomes the norm binding us to the world. In Heidegger's terms, the exhibition, because of its dominance and prevalence, becomes the norm. Heidegger's (1977, p. 81) point is that the world in the picture "stands before us as it stands with it for us," showing that beyond the attraction, people have approached the world through the picture. How, in this light, does visiting an attraction become necessary if the world has already been taken as a picture? If people are already perceiving the world through a picture, what is so extraordinary about encountering an attraction and photographing it with a camera?

Existing studies on the experiences of tourists have considered the attraction a major source for going beyond everyday life, particularly the work sphere (MacCannell, 1976; Urry & Larsen, 2011). It is believed that tourists are drawn to attractions because they are seen as spaces where detachment from work and mundane concerns can enhance tourists' well-being and satisfaction in life (Chen et al., 2013; Christou, 2020; Smith & Diekmann, 2017). Idealized as a period of absence (De Bloom et al., 2010), attractions essentially represent the nonwork, or what work is not. However, the central role that the tourist occupies in modern life, such as the "modern man in general" (MacCannell, 1976, p. 1), suggests that the tourist–attraction relationship can fundamentally inform how the modern being exists in the world. Here, I suggest that the attraction's relationship to the nonwork sphere should not be reduced to the dichotomy between work and nonwork, but where tourists can reconsider their working relationship and primary engagement with the world through the attraction. In this light, the attraction provides a perspective on working relationships and the mundane that permeates the everyday life of the individual.

The relationship between tourists and attractions is, as Mitchell (1991, p. 29) describes, a "careful and curious construction". Attractions draw tourists by offering both internal and external experiences. The attraction, as a framed and ordered representation of the world, connects with the tourist through their gaze. The attraction and the tourist occupy different places, similar to how the attraction is a mere representation, not the real world. Furthermore, tourists are drawn to an attraction because the enframing technique establishes a unique position for them. The attraction intentionally surrounds tourists with a complete view of the represented world, creating the illusion that they are at the center of the experience. Yet, tourists remain detached from the attraction, as passive observers, viewing the image without becoming part of it. This curious "participant–observer" position, as Mitchell (1991, p. 28) explains, shapes the relationship between the tourists and the attraction.

While the world-as-exhibition introduces a new method of engaging with the world, Mitchell (1991, p. 29) argues that "it was not just the particular position that was new; it was the very effect of having a position". He emphasizes that the possibility of adopting the perspective of a participant-observer is a powerful and pervasive experience, making tourism an inevitable part of modern life. In their everyday lives, individuals are conditioned to view the world as a series of images, while attractions provide tourists with an opportunity to reflect on how they have perceived the world as images. In other words, being drawn to an attraction is both a response to and an affirmation of one's position before the world on display. This space for reflection – looking at one's own act of looking within a world picture – gives the attraction its distinct and valuable role as a commodity.

One of the most prominent examples of the participant-observer role is Jeremy Bentham's inspector in the Panopticon. The core idea of the 19th-century English philosopher is that the Panopticon embodies the principles of exhibition and techniques of observation (Bentham, 2020). The inspector, positioned centrally in the Panopticon, can observe both the prison and its inmates from a hidden, dark spot. Like Mitchell's tourist viewing the attraction, the inspector is both inside and outside the exhibited prison. Bentham (2020) concludes that the inspector uses surveillance techniques to ensure the prison and prisoners are arranged in an acceptable order. Hollinshead (1999, p. 7) draws a parallel between the tourist and the inspector, stating that the tourist is

> someone who not only participates in the regulation of the world and in the mastery of its social, cultural, natural and geographical environments, but who regulates and thereby constrains himself/herself through the ocularcentric outlooks which he/she upholds.

This dual insider–outsider position – the effect of occupying a touristic position – creates a new lens through which modern individuals can glimpse the world as an exhibition. However, while the Panopticon offers a model illustrating this insider–outsider dynamic, the impact of such a position should not be reduced to its symptoms, such as the creation of attractions or the subjectivity of the tourist. Instead, I argue that the best way to understand the impact of this position is through its visual forms. By visual, I refer to the position that cannot be fully captured through images of a panopticon building but is instead intimately linked to the body and sensory experience of the individual. Foucault (1990) captures this embodied effect of position through the concept of "biopower". Biopower, according to Foucault (1990), refers to a form of power that regulates and controls the population by managing biological and life processes. What is unique about this technique is that it operates through its connection with the human body.

For tourists, this technique mobilizes a particular form of gaze that produces the insider–outsider position unique to the relationship between the tourist and the attraction. The gaze, as a technique, possesses both universal form and transformative power. The gaze connects the public with the object, always transforming the former into a tourist and the latter into an attraction. Developed and refined by institutions like museums and zoos, the gaze, due to its biological relevance, holds the power to regulate and reorder the objects and people that engage with it. Its universality means that any object or person can be processed through the gaze, given sufficient time or space.

The attraction: A Marxist critique

Approaching the look as a form of biopower, or more precisely as a bio-technique, provides a crucial step in distinguishing the attraction from its fetishized allure. By asserting the look as a bio-technique utilized by institutions such as museums and zoos, we are led to a Marxist critique that challenges several assumptions found in existing studies on attraction. To understand this better, we must return to the foundational claim of this section: The attraction is a commodity exchanged for profit in the capitalist market.

A common assertion in discussions of the tourist experience is that the charm of the attraction stems from the observer/outsider position the tourist assumes while engaging with the object of their gaze. This outsider position is often regarded as a privilege, as it allows the tourist to view the exhibited attraction, its history, and tradition from a distance. In MacCannell's (1976) terms, the attraction becomes an "other", something removed from the everyday. For Macdonald (2007), this outsider perspective reflects the politics of colonialism, imperialism, and Orientalism, enabling the tourist to gaze upon the exhibition from a privileged stance. In this way, the attraction, exposed fully before the tourist's gaze, provides the tourist with a sense of mastery and control. This outsider perspective has often been idealized, considered "freer, more innocent, more spontaneous, purer, and truer" (Wang, 1999, p. 360).

While the outsider's viewpoint might be seen as a necessary and desirable trait of the attraction, it is, in fact, the insider position – the position that makes the outsider's view possible – that warrants more attention. While the tourist's gaze allows them to assume an external position relative to the attraction, this look, as a bio-technique, is crucial because it binds the tourist to the attraction. The insider position implies that the tourist is in a relationship with the attraction, and this relationship continues through the act of looking. The idealized externality that the tourist experiences as an outsider is, in reality, defined through their role as an insider. This intimate relationship with the attraction, as a commodity, is analogous to Marx's (2000) conception of people's social relationships in capitalist society: Individuals are mediated and structured through the exchange of commodities.

One of Marx's (2000) primary concerns with people's mediated relationships through commodities is alienation. The bio-technique of the look produces a profound connection with the commodity form of existence. In their close coupling with the tourist attraction, individuals experience the most complete form of alienation within the capitalist framework: In the panopticon tower, the inspector, or the tourist, engages predominantly with the exhibition. They do not see each other in the tower because everyone is absorbed by the exhibit. In this sense, they are outsiders – external to any meaningful social relationships. The absoluteness of the gaze ensures their complete detachment from other actors in the system including capitalists who control the techniques of the look and the means of production, fellow tourists similarly entranced by the exhibit, and even their own bodies, confined in the inspection tower. It is in this extreme solitude and individuality that the tourist encounters the illusion of externality and outsider.

As a fully realized commodity, the attraction and its "nonwork" denotation do not escape the capitalist system. The relationship the tourist forms with work – through the display of work – should not be idealized as an antithesis to labor. Instead, the tourist's experience with the attraction maintains the separation between labor and the means of production, which Marx (2000) identifies as a cornerstone of the capitalist system. The bio-technique of the look serves to witness this separation in its purest form. The attraction, as a commodified exhibit, both separates the museum or zoo as an accumulated labor process from the tourists and transforms this separation into a necessary social process through commodification. As Marx (2000, p. 522) notes, "as soon as capitalism is firmly established, it not only maintains this separation but reproduces it on an ever-expanding scale". The tourist attraction, both symbolically and materially, reproduces this separation, thereby maintaining the foundation of capitalism through its carefully crafted structure.

The giant panda attracts

This chapter begins with a question: Why did the giant panda captivate millions of Western audiences in the 1930s, while her obvious allure seemed lost on the Chinese? Viewed through a Marxist lens, the giant panda's natural charm reflects a fetishistic quality attributed to commodities by her consumers. Western tourists' enthusiastic embrace of the giant panda as an attraction was shaped by the capitalist social structures available in the Western world that commodified her. By contrast, Chinese society at that time remained largely indifferent to the panda's appeal due to the absence of capitalist production cycles and social relations. However, this absence does not mean that the animal, Chinese society, or the local people were excluded from the commodification of the giant panda. Rather, it signifies that the Chinese were incorporated into the capitalist system through an alternative path, where their labor was abstracted to contribute to the panda's commodification. As museums and zoo exhibitions flourished under the themes of colonialism, imperialism, and Orientalism (Macdonald, 2007; Mitchell, 1991), both the Chinese and the panda became subjects of these forces. This process is explored through a story of the giant panda as told by Chiang Yee.

Chiang Yee (1903–1977), born in Jiujiang, China, left his home country for England in 1933, disillusioned with the Nanjing government. Known as *The Silent Traveler*, Chiang's artistic talents gained recognition through his *Silent Traveller* series, which detailed his experiences in Western society. These works earned him acclaim among British intellectuals and artists, and he was even nominated for the Nobel Prize in Literature in 1973. Chiang also authored several children's books, with the giant panda as a recurring protagonist. *The Story of Ming*, published in 1947, was inspired by Ming, the first panda cub to arrive in the United Kingdom in 1938.

In *The Story of Ming* and other panda-themed children's stories, Chiang presents pandas as animated, sentient beings capable of thinking and feeling. I argue that Chiang's use of children's literature to narrate the giant panda's experiences aligns with local Chinese efforts to mythologize the panda, as discussed in Chapter 1. This approach served as a means to address crises such as British racism, Western imperialism, and Chiang's homesickness for China – issues he addressed in his other works. Furthermore, Krueger and Krueger (2018, p. 127) observe that "an illustrated children's book, with talking animals, must tell the truth, resonate emotionally, and graphically depict concerns". Narrating the transformation of an animal into an attraction from the animal's own perspective, it seems, was well-suited to the medium of children's stories.

The Story of Ming begins with the birth of a baby panda in a loving panda family. The young panda enjoyed her family's care and played with her human friend, Chin-Pao, until one day when the family encountered a Western explorer on a trekking path. This meeting marked a turning point, fundamentally altering the baby panda's life as she embarked on an adventure to London. Accompanied by her family, the baby panda traveled with the Western explorer to London, where she was warmly welcomed by children. At the zoo, the panda baby received her Chinese name, "Ming".

Although written as a children's book, Chiang's *The Story of Ming* captures critical moments in the panda's transformation into a London zoo attraction. The Western explorer in the story is modeled after Floyd Tangier Smith, a Japanese-born American adventurer who collected specimens for American institutions in East Asia. The encounter between Tangier Smith and the panda family represents the first pivotal moment in the baby panda's journey to becoming an attraction. Chiang illustrates this transformative moment through both text and imagery.

In the lower-right corner of the illustration, Tangier Smith, dressed in a gray suit and holding a trekking pole, observes the climbing giant panda cub from a vantage point atop a rock. In the upper-left corner, another rock hosts five pandas playing with a Chinese boy. The illustration subtly contrasts the Eastern and Western figures – Tangier Smith's image is one of masculinity, adulthood, dominance, activity, and strength, while the Chinese boy appears passive, childish, half-sleeping, and unconscious. However, the key difference lies in the interaction between the Western figure and the giant panda, as opposed to the Chinese boy's. Tangier Smith's intense gaze is focused on the panda cub, a dominant feature of the image. Despite the Chinese boy being physically surrounded by five pandas, it is Tangier Smith's eye contact with the panda that commands the viewer's attention. Positioned closer to the reader, the Western figure towers over the panda, asserting an intense, touristic role. In this gaze, the giant panda seems caught, immobilized under the scrutiny of the Western observer's technical and investigative look.

Next to the illustration, Chiang (1945) offered the following narration for this encounter:

> It was one of their expeditions with Chinpao that all five giant Pandas came across an explorer from the West. He was in the mountains of China collecting various things for museums in England and America. It was the first time he had seen this animal with black ears, He was very interested and thought that his fellowmen, and especially children, would be interested too. He decided to take all five of them with him to Europe. He was prepared to take Chinpao with

> him too if the boy's elders would let him go. At first, father and mother Panda made all sorts of difficulties and flatly refused to leave the mountain. Uncle and aunt Panda supported them. But the baby was terribly excited and longed to set out without delay. They could not resist her enthusiasm and certainly they could not let her set out alone on such an adventure, so at last they all agreed that the whole family should accompany the explorer. Only Chinpao did not go with them.

Tangier Smith considered the giant panda of great interest, a fascination that allowed him to possess the animal as his private property almost instantly. He reflected on the need of his fellow men and children to see the giant panda, shaped by his understanding of the commodity of tourist attraction and the desires of consumers. Tangier Smith established his ownership of the giant panda without hindrance: A mere encounter in the mountains was enough to change his relationship with the panda and the life of the animal. In contrast, the panda's playmate, the Chinese boy Chinpao, did not possess the animal. While Chinpao was considered the panda's companion, Tangier Smith was prepared to take him to the United Kingdom.

Tangier Smith's reflection that Europeans would be romantically interested in the animal served to aestheticize his commodity-exchange-based relationship with the public. His understanding of European interest in the giant panda demonstrated that the explorer connected with the Western consumers through mediated commodity exchange. This interest quickly labeled the giant panda into a rare and significant means of production from which Tangier Smith could profit.

However, what is particularly interesting is Chiang's observation of the panda family's reaction to their encounter with Tangier Smith. In Chiang's story, the panda family was torn between two poles: The older generation – father, mother, uncle, and aunt – all refused to leave the mountain in the face of undue difficulties. On the other hand, the panda baby was excited and drawn to her adventurous future, "longing to set out without delay". Eventually, the panda family compromised, swayed by the panda baby's enthusiasm, and decided to leave the mountain with the explorer. What could be so exciting about becoming an attraction and a commodity for the giant panda baby? If turning into an attraction means objectification and consumption, was the panda baby simply idealizing and fantasizing about her journey with Tangier Smith?

The Marxist scholar Louis Althusser (2006) conceives this paradox as an ideology that "hails" or interpellates individuals as concrete subjects through the operation of the subject category. In this process of interpellation, ideological systems call individuals into specific social roles or identities. In everyday life, this interpellation is akin to the police calling out, "Hey, you there!" Interpellation constructs social relations for individuals and places them within the larger social structure. Judith Butler (2002), in *Gender Trouble*, suggests that gender relations also interpellate one to become a woman. Butler emphasizes that the interpellation of a hail compels the individual to respond. The experience of pleasure is one of the conceivable focuses of interpellation because "which pleasures shall live and which shall die is often a matter of which serve the legitimating practices of identity formation within the matrix of gender norms" (Butler, 2002, p. 90).

Contextualizing the panda baby's excitement and longing to leave with Tangier Smith in the process of interpellation shows that the panda identifies with her incoming transformation into an attraction. The legitimacy of this identity formation is her experience of pleasure and excitement through this transformation. Tangier Smith's gaze hailed the giant panda to become an attraction, and the baby panda, in turn, was compelled to accept this role. Butler (2002) demonstrates that interpellation can also manipulate and construct the experience of pleasure or displeasure. Instead of asking why becoming an attraction was exciting and pleasant for the panda baby, her excitement and desire to leave are consequences of her identification with the role of attraction. Responding to Tangier Smith's hail afforded the panda baby pleasure and excitement – experiences she would not have had if she had stayed with her family.

The panda family's arrival in London, however, turned out to be less of an excitement for the panda baby:

> They soon found themselves surrounded by much bigger crowds of people than in Chungking. The baby Panda decided that London children were almost too affectionate and excitable—they were always wanting to hug her and stroke her. This she did not like at all, but she put up with it patiently and good-naturedly.

The identification with an attraction incorporated the panda baby into a new set of social relationships where she was continuously gestured at and pointed to – London children petted, hugged, and stroked her. Although the giant panda baby did not enjoy these experiences, she was firmly bound to her role through her patience and good nature. It is important to recall that the panda baby had been a disruptive, restless, rebellious, and fierce figure during her family's adventure. Her enthusiasm and impatience to leave her home had, as we were told, been the driving force behind the family's compromise to become the attraction. Here, we must recontextualize the panda family within their Chinese cultural framework to understand how difficult and almost unlikely this decision was for the panda family.

Governed by the Confucian virtue of filial piety, the Chinese family is traditionally organized in a hierarchical structure that prioritizes the family's interests over individual desires (Han, 2012; Ho et al., 2012). Respect for elders, alongside male and senior dominance, is the foundation of this structure, making individual desires secondary to family obligations. In this context, the panda baby's desire to leave her home was a direct challenge to the wishes of her senior family members, who wanted to remain in the mountain rather than venture into the unknown London. Yet, the panda family compromised, placing the panda baby's self-interest and desires above the family's filial piety, the tradition that had historically held the family together. Was the panda family's decision simply naïve or manipulated because they indulged their youngest member's self-interest? Did life in Europe hold such tremendous allure that even a traditional family like the pandas could not resist?

In Chow's (2003) study of her mother's reflection on Bernardo Bertolucci's *The Last Emperor* – "It is remarkable that a foreign devil should be able to make a film like this about China. I'd say, he did a good job" – Chow suggests that ethnic Chinese identity experiences a successful suturing between an Orientalized image on the screen and identification with the ethnic other. This ethnic spectator occupies an impossible space, the continuity of which depends on her "engagement with what is culturally available" (Chow, 2003, p. 24). I propose that it is possible to view the panda family's compromise and dilemma in a similar light, as an impossible space whose existence can be justified through the capitalist terms. The panda baby, as the story has demonstrated, identified with her commodification and attraction form before being rewarded with the pleasure and excitement that such identification brings. Therefore, her desire to leave was not merely an expression of self-interest or longing but the incorporation of capitalist relations into a traditional cultural system that could only respond to what was economically available. The panda family did not abandon traditional Chinese family values for the sake of civilized European culture. Instead, their decision was calculated and carefully, though painfully, measured, recognizing that commodity and commodified relations would prevail.

The panda baby's transformation into an attraction for children in London also marked the dissolution of the traditional panda family bond. The disappearance of this bond, in the face of the attraction, highlighted the panda baby's individual existence. The panda family, which had once borne and supported the panda baby's life, could no longer protect her from facing the overly affectionate and passionate audiences. Since her dislocation, both physically and emotionally, from her home and family, the panda baby experienced what Fanon (2023**, p. 18**) describes as an "inferiority complex". Was life in London necessarily better or more exciting than her life with her family and Chinpao in the mountains? Where would the panda baby stay, given her current situation?

The narrative offers no clear indication of the panda baby's transition from an enthusiastic, easily excitable, and hot-tempered individual to a patient and good-natured being. Instead, we see that, in this state of nowhere, uncertainty, and inferiority, the panda baby nonetheless received her Chinese identity:

> It was at the London Zoo that she was given by the Zoo authorities the name of Ming. "Ming" is a Chinese word, written in Chinese like this 日月 or like 明, a combination of two parts. As both sun and moon give light, the world "Ming" means "bright" or "brightness." It is a very good name for the baby Panda, as it tells that she comes from China and suggests that her mind is always full of brightness in spite of her half-sad face.
>
> *(Chiang, 1945)*

Who was the panda baby before the London Zoo gave her the name *Ming*? How did the panda family address her? Did she have a name from her parents? The story suggests that who the panda baby was before her move to London and how her family had addressed her were not important for her transformation into an attraction, nor significant for the London visitors who came to see her. What mattered at that moment was that the panda baby was "Ming", an animal the London Zoo could identify as Chinese.

Anyone familiar with the Chinese language would understand that the name *Ming* could hardly be considered a typical name for children in a Chinese family. The character *Ming* (明), standing alone, often refers to the Ming dynasty (1368–1644), a historical imperial period in China that predates the fall of the Qing dynasty. This connection to a distant, fallen dynasty, regardless of the character's literal meaning of "brightness", makes it an awkward choice for a child's name. Yet, when taken out of its cultural context, the name *Ming* seemed a fitting designation for the giant panda, far removed from her habitat. The irony is that, although this dissection of Chinese culture may appear awkward to a Chinese audience, the name *Ming* turned out to be "a very good name for the baby panda" (Chiang, 1945).

This tightly intertwined experience of being both a giant panda and Chinese, through the lens of becoming an attraction for Western audiences, suggests that Chiang's story is about both the animal and the Chinese at the same time. Is it a story about the Chinese through the image of the giant panda, or a story about the giant panda through her Chineseness? This ambiguity of speaking through an animal to express one's native culture encapsulates the experience of dislocation and alienation that the bio-technique of attraction enforces in the creation of the giant panda attraction. Scholars like Gayatri Spivak (2003) have explained the difficulty, if not the impossibility, of speaking from and about the experience of becoming the other – in this case, the giant panda attraction. John Berger (2015) notes that this ambiguity, or impossibility of speaking, can also influence the perspective of the tourist because "in the zoo the view is always wrong". The wrong view, as Berger argues, is that the animal never returns the look at the tourist, observing that "nowhere in a zoo can a stranger encounter the look of an animal". For Berger, the zoo's representation of animals and attractions are irredeemable within capitalist relations – the panda attraction can only tell the story that the London Zoo gives, and never the story of her abandoned mountains and family back in China.

However, it seems that there is no better place to begin addressing what is wrong with the zoo attraction. Is there a different way to confront what is deemed legitimate – the right look of the tourist – through what is meant to be wrong? If the zoo attraction is always wrong because the animals refuse to look back, do they also tell us that something has not been looked at, transformed into an attraction, and commodified? If there is part of the animal that rejects the order of commodity and disrupts the individual–commodity entanglement, should we take a different look at how animals look – or fail to look – at us? I am not idealizing animal consciousness or the body as being beyond the reach of contemporary society or social relations, nor am I asserting that animals represent a realm of primitive passion and undomesticated power (Kristeva, 2024). As Foucault (1990) reminds us, power and resistance always coexist within a complex network. It is reasonable to assume that within the giant panda's attraction, there are traces of the animal's effort to challenge or subvert the attraction itself. This prompts us to step closer to investigating the giant panda attraction, rather than distancing ourselves from this construct simply because of its commodified, fetishistic, and superfluous nature. To do this, I turn to the aesthetic effect of cute in the giant panda attraction.

3

A COST-SAVING PANDA IS CUTE

An animal facing extinction because of relentless capitalist expansion and industrial exploitation, the giant panda has nevertheless survived "because it is cute" (Songster, 2018). If we agree with Daniel Harris (2000) and Sianne Ngai (2000) that cuteness, as an aesthetic category, originates in mass consumer culture in the industrial 19th-century United States, explaining the giant panda's survival through her cuteness raises a further question: How does an aesthetic category originated in commodities save animals being exploited precisely for the reason of capitalist commodity production? This question juxtaposes capitalism and the giant panda in its two most unlikely terms: The exploitation of the giant panda, on the one hand, and the rescuing of the giant panda, on the other. The question considers the giant panda as an almost unlikely heroine of the capitalist reign – the animal, super-exploited, does not go extinct but continues to live among us, fully alive.

Is this the giant panda's triumph or the capitalists'? While Marx's idea of capitalism gives us insight into how the animal suffers from her exclusion and double exploitation as a means of production external to capitalist relations, how could we conceive the animal's survival through, also, the capitalist means of consumption? Considering that the cuteness of the giant panda is the neutralized and naturalized fetishism of the commodity for attraction, does it suggest that commodity fetishism, which Marx worries about because it denounces labor performed by humans, can, in turn, be helpful and even righteous for animals such as the giant panda?

One of the classical stories circulating among senior panda researchers to junior researchers is the creation of the landmark World Wildlife Fund (WWF) logo. Often, the story begins with Sir Peter Scott's inspiration, the famous giant panda Chi-Chi, who resided in London Zoo since her arrival in 1961. In the story, Scott was moved by the animal's beauty and appeal. The narrator then impresses the audience with a twist: "But you know what? A more important reason for using the

DOI: 10.4324/9781003511229-4

giant panda as the logo was to save money because the panda is black and white!" To add to the credibility of the story, Scott's remarks on the official WWF website (2022) are often cited: "We wanted an animal that is beautiful, is endangered, and one loved by many people in the world for its appealing qualities. We also wanted an animal that had an impact in black and white to save money on printing costs". While the story is often laughed off for its humorous and unexpected turn – given that the WWF logo is now one of the most valuable trademarks in the world today (Barua, 2020) – this economic-saving aspect of the giant panda deserves further examination. Should we describe the giant panda as a cost-saving animal? Or the giant panda as an impactful animal? An animal that reduces printing costs?

Somehow this cost-saving observation had only added to the cuteness and aesthetics of the giant panda. In contrast, it seems that the category of the beautiful is less capable of capturing this situation of the giant panda. Beauty and the appreciation of nature, since Kant (1952), after all, are intimately connected with moral character. And what is the moral for a money-saving animal? Surely the giant panda in Peter Scott's view is not sublime – the animal does not evoke feelings of awe, respect, astonishment, or transcendence as a response to nature's powerful, vast, and overwhelming presence (Kant, 1952). Instead, the animal is tied closely to a relationship with a different industrial technology – the printing press – and Marx's exchange value, the price/cost of the print. The cuteness of the giant panda, it seems, does not only have to do with her appealing appearance but also with how the animal helped WWF save costs. Indeed, is not a money-saving giant panda, by all means, cute?

Ngai (2010, p. 1) considers cuteness, one of the aesthetic categories indexing and materializing the capitalist processes of production, circulation, and consumption, as the current "most socially binding process". Hence, as an aesthetic category specifically relevant to the conditions of late capitalism, cuteness also calls forth "not only specific subjective capacities for feeling and acting but also specific ways of relating to other subjects and the larger social arrangements these ways of relating presuppose" (Ngai, 2012, p. 11). In Ngai's (2012) terms, the cuteness of the giant panda draws on people's relationships in capitalist society and maps a collective response shared between people attracted to the giant panda. The first section of this chapter follows this theoretical strand to further investigate the giant panda as a means of production in the attraction and its commodity form. The aim is to show how cuteness, as an aesthetic category and response, can inform our understanding of the super-exploited status of the giant panda in capitalist relations.

The cultural cuteness

Existing studies have examined the concept of cuteness and its aesthetic effects through two primary lenses. The first traces the experience of cuteness to biological and genetic responses triggered by encountering animals or objects displaying infant-like characteristics. Seen as a natural instinct embedded in human genes, the experience of cuteness is thus considered an innate response. The second lens links cuteness and its aesthetic appeal to the rise of commercialism in 19th-century industrial America, where it played a role in ideologically consolidating the feminization of the middle-class home, organizing this space through commodities and consumption. Cuteness, in this view, becomes a cultivated aesthetic experience intertwined with the capitalist processes of production, consumption, and circulation.

In 1877, Charles Darwin published his influential note “A Biographical Sketch of an Infant” in the journal *Mind*, suggesting that infants’ emerging behaviors could be deemed instinctive and innate. Darwin (1877, p. 286) observed that behaviors like winking and wrinkling the face “manifestly serve to protect them” and were “not acquired through experience”. In his close observations of his infant son, Darwin (1877, p. 288) noted that smiles, which appeared around 45 days old, were “probably of mental origin”, especially when directed at the mother. Lorch and Hellal (2010) argue that this article helped establish the “natural science of babies”, providing evidence with significant implications for Darwin’s broader evolutionary theory, which was not widely accepted at the time. Contextualized within Darwin’s theory of evolution, the infant–parent connection becomes a product of natural selection and the struggle for survival. Despite their relative immaturity compared to other primates, human infants have evolved survival strategies that elicit care and bonding from parents, a connection Darwin believed was rooted in the origins of the human mind.

This infant–adult connection was formally analyzed and discussed when Konrad Lorenz, the Austrian ethologist and Nobel laureate in 1973, conceived the notion of *Kindchenschema*, or “child schema”. In *Kindchenschema*, Lorenz identifies the features that trigger an adult’s investment in infants: “a relatively large head, predominance of the brain capsule, large and low-lying eyes, bulging cheek region, short and thick extremities, a springy elastic consistency, and clumsy movements” (Lorenz, 1971, p. 154). Lorenz (1971, p. 154) explains that “following the law of summation and combining”, these features, when combined, give the child a “loveable or ‘cuddly’ appearance”. In a figure illustrating the releasing schema for human parental care responses, Lorenz (1971) paralleled images of four examples – child–man, jerboa–hare, Pekingese dog–hound, and robin–golden oriole – showing how the cuddly features should be captured through images of the former.

Lorenz’s idea of *Kindchenschema* has since inspired follow-up studies (Cho et al., 2022; Dale, 2017; Doebel et al., 2022; Dydynski, 2020) that have modeled and elaborated on the concept of cuteness using these terms. Despite the different scientific aims and contexts of these studies, I want to draw a parallel between Lorenz’s observation of cuteness and Armand David’s encounter with the giant panda. It is perhaps by comparing these two contrasting and seemingly unrelated cases that we can uncover the resonance between the observation of the giant panda and the concept of cuteness. In Chapter 1, we noted that David’s encounter with the giant panda was regarded as a scientific discovery that introduced the image of the giant panda to the scientific community. However, this encounter does not imply that the giant panda did not exist before: The animal had roamed the mountains of Sichuan for centuries. David's encounter with the panda, underpinned by imperialist expansion, colonial agendas, and the capitalist system, marked the animal as a reliable resource for further Western exploration.

What does Lorenz's *Kindchenschema* have to do with David's encounter with the giant panda? It seems that both scientific discoveries followed a similar trajectory. Lorenz's observation of *Kindchenschema*, or what we might call the "scientific discovery of the cute", was not the invention of the cute human baby per se. The human baby, like the giant panda, has always existed. Should we argue that the human baby, before Lorenz's observation, was not cute at all? What made Lorenz's scientific discovery significant was that he ensured the concept of cuteness could be visualized – captured through a set of specified features. By detailing characteristics such as a large head, protruding forehead, big eyes, short chin, thick extremities, and clumsy body movements, Lorenz provided a framework through which a "cute" human infantile image could be conceived and grasped.

It is Lorenz's (1971, p. 154) further reflection that the cute features follow "the law of summation and combining", which, although it has received limited attention from his fellow scholars, I believe completes Lorenz's description of the cute. Despite Lorenz not explaining what the law of summation entails, he made it clear that cuteness and its image are an arranged, selected, and controlled representation based on the features of the human infant. This suggests that not every infantile feature of the human baby is considered cute, but rather a specific set of identifiable characteristics. In this light, Heidegger's (1977, p. 81) concept of the world picture – "the world conceived and grasped as picture" – can provide a footnote to Lorenz's law of summation and combining. In other words, what appears to be cute is presented as such because a particular way of seeing, connecting, and collecting has identified the similarity between the human baby and the object of cuteness. Viewing Lorenz's *Kindchenschema* through Heidegger's idea of the world picture has at least two significant implications.

On the one hand, the cute and its representation are conceived and arranged through the image of the human baby. This means that cuteness and its collected features are anthropocentric. Perceiving any object as cute amounts to an aesthetic judgment, with the human infant as the yardstick. Heidegger's (1977) concern with the world picture and the establishment of man's subjectivity through the viewer's perspective is that human subjectivity – one's view and experience – takes precedence over other centers of reference. Cute animals, such as the jerboa, pekingese dog, and robin, are made visible because they share certain features with the human infant. In contrast, adult animals that fail to exhibit these infantile features are disqualified from being considered cute. The giant panda, a nonhuman species that exemplifies the image of cuteness, demonstrates a collection of these human baby features, including large eyes (despite the impact of their black patches), fluffy bodies, clumsy movements, large heads, and pug noses. Despite their large size, adult giant pandas are still considered cute because adulthood does not severely interfere with these features.

On the other hand, the viewer's experience with the cute is that of a participant-observer, or what Mitchell (1991, p. 21) calls an "objective person". In this dual position – both external and internal to the image of the cute – the viewer experiences detachment and attraction to the cute object. This objective viewpoint is critical to the making of modern human subjectivity because, as Heidegger (1977, p. 84) asserts, "the position of man is conceived as a world view". This particular way of looking at the world through a constructed and conceived image produces contemporary human subjects. This situates the concept of cuteness within its social–material context, suggesting that cuteness is but one of many worldviews collectively laboring to shape human subjectivity. In the case of David's scientific discovery of the giant panda, we have already seen how a single discovery mobilized ongoing West–China interactions and brought the giant panda to the global stage. Suffice it to say that the transformation of the giant panda into a world image seamlessly integrated the animal into the forces of global capitalism. We can observe this transformation because the giant panda's world picture, due to its historicity, allows us to reflect on and "look back" (to use Heidegger's term). The same can be said of cuteness, although we need to take a different step in considering the collective impact underlying cuteness, given its prevalence and ubiquity in everyday life.

It can be argued that images perceived through the world view, according to Heidegger (1977), can serve as a means to produce human subjectivity, meaning that cuteness is one such means. How, then, does cuteness, together with its widespread presence, further shape human subjectivity? For the cultural critic Harris (2000), cuteness generates a deception in everyday life and consumerism. Harris (2000, p. 2) notes that consumers' "conditioned responses to this most rigid of styles prevent them from recognizing its artificiality" because "the very embodiment of innocence represents an absence of the designed and manipulated qualities of what is, in fact, a heavily mannered aesthetic". The human subject shaped by this deceptive cuteness is driven by a sadistic motive, as the cute encourages human abuse and violence through its innocent and vulnerable appearance. Many studies echo Harris's concerns about cuteness, suggesting that when confronted with an image of the cute, the audience is provoked to respond through their parental instincts, and thus, the cute and its objects are necessarily projected as feminine (Merish, 1996; Ngai, 2005), powerless (Merish, 1996; Ngai, 2005), dependent (Feng et al., 2022; Zhang & Zhou, 2020), and "small things" (Feng et al., 2022; Zhang & Zhou, 2020). While the audience experiences feelings such as pity, empathy, and compassion for the cute object, they also experience sadistic pleasure in exploiting the object's obedience and vulnerability. As Harris (2000, p. 6) puts it, "the process of conveying cuteness to the viewer disempowers its objects, forcing them into ridiculous situations and making them appear more ignorant and vulnerable than they really are". The capitalist market capitalizes on this instinctive tendency to sympathize with and parent the cute, generating profits through the image of the disempowered, distorted, helpless, and deformed cute products.

More recent studies have attempted to show that consumers' relationships with cute objects go beyond sadistic violence and are full of complexity (Dale et al., 2016; Harris, 2000; May, 2019; Merish, 1996; Ngai, 2012). Dale (2017), for example, proposes that cuteness contains a "dark side", namely, the fantasy that allows the cute object to dominate, control, and manipulate its audience. Hence, Dale (2017) points out that the experience of cuteness is masochistic. Likewise, Ngai (2012) shows that cuteness, because of the physical intimacy it invokes between people and the object, is a way of resisting the logic of commodification. Ngai (2012, p. 12) suggests that the cute has a "certain utopian edge" that the capitalist market logic cannot easily manipulate. This gives cuteness an ambiguity that allows it to both control and be controlled. In *The Power of Cute*, Simon May (2019) agrees with Ngai's point, noting that the cute "lacks clear identities" (40) and suggesting that the cute one's "surrender is a way of entrapping her (the audience)" (45). Precisely because of this ambiguity and uncanniness, the cute, according to May (2019, p. 48), is amoral and alien to responsibility – for this space is neither moral nor immoral. This allows the cute to be powerful because it aestheticizes the contemporary experience of human life and has been successful in consolidating this experience.

The biological cuteness

While I agree that cuteness is an aesthetic category prevalent in capitalist relations, a point I will return to later, I believe that to justify the concept of the "cute" in Heidegger's (1977) terms – namely, a worldview that conceives and arranges the world – it is necessary to probe the questions of how and why: How is cuteness formed? And why is it so? This requires not only a return to the presence of cuteness in capitalism but also an exploration of its original link to scientific interest, first introduced through research on human babies. Ngai (2012, p. 1) has suggested that cuteness emerges in line with other aesthetic categories within the "hypercommodified, information-saturated, performance-driven conditions of late capitalism". From this perspective, cuteness is largely seen as an aesthetic response to the progression of capitalism, and the contradictory feelings toward it (sadistic–masochistic, aggression–sympathy, and care–violence) are readily contained within individuals' experiences with commodities. However, these feelings do not only emerge in interactions between people and commodities. Rather, the instinctive responses to offer parental care and the interest in human infants have been essential to the survival and history of the human species. The fact that cuteness taps into these primitive instincts and emotions, and further links them to capitalist production and circulation, could be addressed through Foucault's framework of biopower – how modern societies govern and regulate populations through techniques that focus on life processes.

In this context, Darwin's theory of evolution becomes not only relevant to but also central to the concept of cuteness and its aesthetic impact. Darwin's interest in human child development was integral to his theory of human evolution. The child, as observed by Darwin in his own son and daughter, was examined closely to provide evidence for his evolutionary theory, while Darwin is often remembered as one of the most loving and caring fathers among scientific figures. The publication of "A Biographical Sketch of an Infant" marked a critical moment in consolidating the acceptance of the theory of evolution. As Lorch and Hellal (2010) note, this article foregrounded the "natural science of babies".

Simply put, the idea of evolution refers to the alteration of inherited traits within biological populations across successive generations. Evolution consists of two independent components: Randomness (or genetic drift/alteration) and natural selection, which is passed down through generations. Randomness is inconstant and capricious, and cannot create anything organized. Natural selection, on the other hand, promotes the integration of beneficial and adaptive traits into the design of a species over the course of evolution (Dawkins, 1996). Darwin's idea of evolution provides a framework for understanding a species' cognitive or neural structures as the result of the collaboration between randomness and natural selection. These consequences can be categorized as adaptations, byproducts, or genetic noise (Dawkins, 1996). Adaptations are traits that were favored by natural selection for their role in promoting genetic propagation. Byproducts have no inherent function but are present because they are causally linked to adaptive traits. Genetic noise arises from the random or stochastic elements of the evolutionary process.

The human infant's presence, both mental and physical, contextualized within evolution, is the result of gradual adaptation through heritable characteristics and the selection of these alterations (Hall & Hallgrímsson, 2008). One of Darwin's younger contemporaries, James Sully (cited in Lorch & Hellal, 2010, p. 150), suggested the following:

> The doctrine of evolution bids us view the unfolding of a human intelligence today as conditioned and prepared by long ages of human experience, and still longer cycles of animal experience . . . . [A]ccording to this view the successive stages of the mental life of the individual roughly answer to the periods of this extensive process of organisation-vegetal, animal, human, civilised life . . . . [I]t is in infancy that we are least indebted to our individual exertions, mental as well as bodily, and that our debts to our progenitors seems heaviest.

What Sully suggested affirms Darwin's assertion that human infants possess inherited traits that enable them to survive, and these traits seem particularly "pronounced" because they are adaptive. For both Darwin and Sully, the infant must be seen as a collective and prominent expression of the adaptations the human species has evolved over time. For example, Darwin (1877) links a baby's smile at its mother to the origin of the human mind. Similarly, the philosopher Kimerer

LaMothe (2015, p. 131), when observing her newborn son smile at her, pondered similar questions:

> Rationally speaking, his smiling doesn't make much sense. There are so many other tasks he must master before he can fend for himself. He cannot yet roll over, sit up, crawl forward, or wield a spoon. Why spend any energy at all learning to smile? Doing so involves activating at least forty-four muscles in the face. Shouldn't he be devoting that brain space to something critical to his survival? Or could it be that smiling is? Could it be that smiling is, for him, a life-enabling pattern of movement?. . . When Leif cracks a smile, I fall in. Maybe that is its purpose—to catch me, a caregiver. Maybe I am so moved because I know what it means for me to smile.

LaMothe reached the same conclusion as Darwin that the baby's smile can make a life-or-death difference to her and the smile originates in the human mind. How, then, should we approach Lorenz's *Kindchenschema*, a set of human infant cues understood as cute, in relation to the human species' adaptive strategies? In the context of evolution, the cute appearance is not a mere accident but an inherited and carefully selected trait that will maximize the species' chance of survival. Like the smile that taps into a heartfelt response of LaMothe toward her baby, the cute appearance – big eyes, large head, short chin, awkward movement, or short extremities – connects the baby to their caretakers as an irresistible attraction. We might also venture to ask why the human baby also "wastes her energy" in terms of being cute – would not a fitting head, agile and long extremities, small eyes, flat cheek region, and tactful movement help the baby survive in her environment better? Why bother to be cute, a comparatively helpless and vulnerable status at all?

About two decades ago, evolutionary psychologists such as Tooby and Cosmides (2001, p. 15) asserted that aesthetic experience is a "necessary guidance system for the development of each adaptation", even for activities that often appear nonfunctional or extravagantly nonutilitarian. They suggested that the appreciation of aesthetics is an essential and designed experience within the framework of evolutionary theory. This acknowledgment of the critical role aesthetic experiences play in evolution suggests that the aesthetic appreciation of cuteness and the human infant's adaptive selection for a cute appearance reinforce each other. On the one hand, the natural selection process has shaped human infants to be cute. Human infants attract adults through all of the senses, making "cuteness one of the most basic and powerful forces shaping our behavior" (Kringelbach et al., 2016, p. 545). Rosenberg (2021) further argues that the cuteness of human infants opens the gateway to human cultural adaptation, as it enhances the infant's capacity to form bonds with their mothers and caregivers, understand the intentions of others, and acquire language and cultural knowledge.

Cute: Biopower and capitalist relations

Noting the critical roles that cuteness plays in the survival and emergence of the human species, it becomes necessary to reconsider its current prevalence in societal, cultural, and economic forms. What does it mean that this life-enabling, adaptive aesthetic experience has become a frequently revisited framework in our everyday lives? The fact that cuteness is an aesthetic experience suggests that its emergence is rooted in dealing with specific aspects of human affection and the senses. Drawing on Darwin's evolutionary theory, it becomes clear that cuteness is connected to life-enabling forces, the survival instinct, the process of creation, and the borderline between life and death. The prevalence of cuteness in the capitalist market – or, as Harris (2000, p. 2) puts it, "everywhere we turn we see cuteness" – suggests that this adaptive strategy, which originally and formally applied to the infant–parent bond, has transformed into a quotidian experience for every individual. While the human infant's survival depends on its cute appearance, as consumers of cute products, individuals can experience a similar bond with commodities. This life-enabling force has been reproduced within capitalist relations, and this transformation of human experience carries significant implications.

Foucault's concept of biopower helps further illustrate the coupling of natural selection instincts and the capitalist system of production in the case of the cute. Michel Foucault introduces the notion of biopower in *The History of Sexuality: An Introduction* (1990) and his lectures *Society Must Be Defended* (2003) as a form of power that takes hold of human life. In addition to disciplinary power, biopower is the second form of power that emerged in modern society, differing from sovereign power. Foucault explains that sovereignty has the "right to kill" and "the right to take life or let live" (Foucault, 2003, pp. 240–241) since the sovereign can kill as part of his dominion over life. Sovereign power "deduces" (Taylor, 2014, p. 41) that it can seize, end, impoverish, or enslave life. What sovereignty cannot seize, it leaves alone, thus constituting a subtractive force, rather than regulating and controlling life (Taylor, 2014). Disciplinary power takes further steps, shaping individuals as objects, goals, and instruments (Hoffman, 2014). The disciplinary power "adds up" through its "meticulous, exhaustive, and continuous control of bodies" (Hoffman, 2014, p. 28). This results in a particular relationship between individuals and disciplinary power, as individuals become both obedient and useful to the power form. Disciplinary power breaks down the group and the whole into individuals, thus not leaving alone life, but instead driving the production of individuals.

In contrast, biopower takes shape in the "right to make live and to let die" (Foucault, 2003, p. 241). While disciplinary power defines spaces and shapes bodies into individuals, biopower works on the living man and transforms him into "man-as-living-being" (Foucault, 2003). As Foucault (2003, pp. 242–243) writes:

> the new technology that is being established is addressed to a multiplicity of men, not to the extent that they are nothing more than their individual bodies,

> but to the extent that they form, on the contrary, a global mass that is affected by overall processes characteristic of birth, death, production, illness, and so on.

While Foucault (1990, p. 139) emphasized that disciplinary power and biopower are two waves of power that emerged in the late 18th century, he notes that they are "the two poles around which the organization of power over life was deployed". Foucault explained through examples like "do not steal" or "do not murder" how the three different forms of power operate (Taylor, 2014). During the period of sovereignty, which prevailed until the end of the 17th century, individuals who committed crimes such as theft or murder were subject to the law and punished solely for their specific offenses. Punishments could involve execution, exile, or fines. With the emergence of disciplinary power in the 18th century, the approach to handling criminals shifted. While individuals were still subjected to the law and punishment, the focus expanded beyond merely the act of crime. The character of the offender, along with their physical and psychological circumstances, became of interest to authorities. This knowledge was crucial for anticipating and preventing reoffending. Consequently, new practices such as psychological examinations, surveillance, and rehabilitative efforts were introduced. Punishment thus became less about ending the criminal's life and more about exerting control through mechanisms like imprisonment, psychiatric treatment, parole, and probation.

In the later 18th century, biopower emerged, shifting the focus of governance toward broader societal trends in crime. Biopower emphasized the overall numbers of thefts and murders within the population, aiming to understand fluctuations in crime rates, identify demographic groups associated with particular crimes, and devise strategies for effectively regulating and controlling crime. While tactics used under disciplinary power were still employed, biopower distinguished itself by focusing on population-level concerns rather than individual cases.

Rabinow and Rose (2006), in their examination of biopower in contemporary society, contextualize biopower through its relationship with authority, discourse, intervention, and subjectivity. They suggest that biopower includes three elements. First, it entails the existence of truth discourses concerning the essential nature of living human beings, with various authorities deemed competent to voice this truth. The truth discourse refers to "a thing of this world, intrinsically bound to apparatuses like the prison, the hospital, the school, and the clinic for its production and circulation" (Rabinow & Rose, 1954, p. 3). Second, biopower involves strategies aimed at intervening in collective existence in the name of life and health. Emerging biosocial collectives, such as categories of race, ethnicity, gender, or social relations, become the target populations. Third, biopower creates individuals who are encouraged to actively work on themselves under specific forms of authority, guided by the truth discourses. Individuals engage in practices of self-improvement or self-regulation, motivated by considerations of personal, familial, or collective well-being. These practices may extend to efforts aimed at enhancing the life or health of the entire population. These shared biological and social identities among new collectives are referred to as "biosociality" (Rabinow, 2022), and specific forms of subjectivity informed by biological discourses and practices are termed "somatic individuality" (Rose, 2009).

Given cuteness's biological roots in evolutionary theory and the survival of the human species, we can now explore how cuteness, as a particular worldview, regulates the global population as a form of biopower.

First, the truth discourse of cuteness is grounded in seminal scientific works such as Darwin's natural science of babies and Lorenz's *Kindchenschema*. The manifestation of cuteness – a summation and collection of human infant features and the parent–infant appeal – is biological knowledge and fact, bound to scientific authority in such a way that it can be conveniently incorporated into commercial terms. Cuteness becomes a taken-for-granted truth that capitalist relations can reproduce on a massive scale for the human population. With the "truth discourse of biology" (Rabinow & Rose, 2006, p. 206) at its core, cuteness allows human beings to replicate a facet of living human beings in their interaction with the world.

In the case of the giant panda, we observe how cuteness becomes a central truth discourse around which the animal has been studied, consumed, experienced, and related to. When Armand David first encountered the panda bear skin in Mr. Li's home in 1869, the French priest commented that the animal was "big and beautiful". He noted the animal's peculiar white-and-black skin and its resemblance to bears. None of the cuteness features – the chubby chin, big forehead, round head, and awkward movement – were highlighted in David's diary. For David, the beauty of the animal was largely dependent on its contribution to the identification of a new scientific species. The concept of "cute" was notably absent in these early encounters with the giant panda. Similarly, for the Roosevelt brothers (1929), the giant panda was "splendid" (226) and "not a savage animal" (263); however, it remained a treasured target of trophy hunters. Sheldon (1937), commenting on their collecting trip for the American Museum of Natural History in 1934, referred to the giant panda as "an extremely stupid beast", "not alert" (14), and "sluggish" (17). Even the panda's remarkable white-and-black skin was not seen as contributing to the animal's appeal; for David, it was uncertain whether "the characteristic markings of the giant panda are protective" (Sheldon, 1937, p. 17). Sheldon concluded his observation with words far from being associated with cuteness:

> I believe further that the animal could be transported to another climate with little trouble. It is dull and primitive in temperament and would not, I think, be sensitive to rough treatment. The only problem, as mentioned before, would be the assurance of an adequate food supply.

Suffice it to say that the cuteness of the giant panda was not immediately recognized in these early panda–human interactions. Rather, the giant panda was judged as beautiful for its contribution to science or dull or splendid for its role in trophy hunting. Barua (2020) points out that, even as the WWF logo since 1961, the giant panda's design evolved to appear cuter over the past four decades. As Nash (cited in Barua, 2020, p. 681) commented, "Sir Peter's panda looked like a panda, but WWF turned it into this thing that looks like a dog". The "appearance of the dog", as we can observe from the transformation of the WWF logo, gradually rounded the outline of the giant panda to cutify its image, emphasizing the big eyes and large head of the animal.

These early documentaries, showing the giant panda's deviation from the notion of cuteness, suggest that the current embodiment of cuteness – or the idea that the giant panda exemplifies cuteness – should not be taken for granted. Rather, it is necessary to highlight the randomness underlying both the giant panda and the concept of cuteness. By randomness, I do not only refer to the genetic drift that could easily direct the evolution of both human beings and the giant panda into different directions but also refer to the selection and establishment of the truth discourse behind cuteness. In the previous section, I noted that *Kindchenschema* was based on Lorenz's visualization of human baby features, and hence made cuteness a possibility through this specific view of the baby. This means that the infantile features of babies were purposefully selected and arranged into the order of cuteness. This makes variant selection or random arrangement highly possible. In other words, what is deemed cute could also mean a set of different or even contrasting features of babies if Lorenz had made his observation differently.

Second, the natural science of babies and *Kindchenschema* were meant to understand the human being as a species. While Darwin was keen to probe the origin of human mind through his observation of the human baby, Lorenz was driven to know more about the behavior of animals and humans and the will of human beings. These biological thinking aimed at treating human beings as a collective existence and wanted to tap into the knowledge of human life and health. The child, the human infant, emerged as the targeted biosocial collective in this research field. Meanwhile, this research agenda is integral to the ongoing economic exploitation that capitalizes on bioscience and mobilizes its elements into new exchange relations (Rabinow & Rose, 2006). In the case of the cuteness, the new human infant features, as the new knowledges of life and health, were being mapped out, developed, and exploited by a range of commercial enterprises (Rabinow & Rose, 2006).

The interest of biopower in collectively managing and controlling the human population through the biological fact of the human species consolidates its anthropocentric approach. With the human population, human life, and human health at the center of the vision of biopower, what is deemed human – the human features, the human mind, the human origin, and human health – provides the yardstick against which the world is measured. Cuteness orients objects toward human baby features and justifies their existence through this "aesthetic kinship". It manifests that objects and things of the world do not always fall into the category of the cute – even human babies themselves can sometimes fail to be cute. However, it is within this unrelenting anthropocentric framework that Harris (2000, p. 3) observes that the cute is "closely linked to the grotesque, the malformed", and an "anatomical disaster". Once distorted or "cutified", the objects can appear as the "charming and the charming grotesque" (May, 2019, p. 27). I believe it is necessary to further argue that this perpetual juxtaposition of the malformed and the charming in cuteness – and hence this apparent ambiguity in the aesthetic expression of cuteness – is not, as May (2019) or Ngai (2012) have suggested, the "essence" or nature of cuteness. Rather, there is a certain aesthetic necessity to which cuteness is bound. It is an aesthetic necessity because of the scientific knowledge and truth upon which cuteness is built – the authoritative and biological fact that the human baby has collective features, such as cuteness. It follows that the aesthetic experience of cuteness always entails malformations, distortions, or deformations that enable the objects and things to appear child- and baby-like. Hence, the aesthetic outline of cuteness will always be a combination of the heavily reshaped grotesque and the charm and appeal of this infantile image.

In the context of biopower, cuteness is not, as May (2019, p. 47) suggests, "amoral, unconcerned with morality", because the cute object escapes blame and responsibility. Instead, the cute object affords no blame or responsibility because cuteness is not about the object in itself but about shaping, arranging, and conceiving the object based on the image of the human infant. In the words of Harris (2000, p. 3), cuteness "is not something we find in our children but something we do to them". Its relevance to the biological fact of human children serves the function of securing power for human beings, as a species, over others and other life forms. This power over other life forms through cuteness can, nevertheless, be captured in an articulate scientific question: Can we drive a life form – the life being – into other things, be they animals, nonhuman beings, the organic world, or commodities, in the image of the cute, since this is how we, the human species, survived in the natural selection process? If cuteness is the selected coping strategy and instinct of human infants for their survival in the harsh natural environment, it is legitimate to assume that cuteness can invoke the same life form in other beings. May (2019, p. 3) observes that cute objects can always "express something richer, and truer to life". What can be richer and truer to life than the power to make life? In this light, cuteness is a form of biopower because it manages and controls the human species based on its understanding of human life and health. The prevalence of cuteness in daily life is a consequence of a consistent sequence of experiments, tests, and the production of life forms on a variety of things in the world. The hailing of an animal or an object as cute is subjecting them to a life performance stereotyped on human beings.

For an animal, the giant panda seems to fit every aesthetic judgment of cuteness. Its similarity to the human infant – or more precisely, the near-perfect visual cues of the animal meeting the criteria of cuteness – has had a decisive impact on the animal's fate. The giant panda survived its potential extinction because of its cuteness. However, this statement must be expanded: The survival of the cute giant panda is itself a manifestation of how biopower manipulates, controls, and manages biosocial collectives at the species level. Because of its similarity to human babies, the giant panda must be loved, saved, and conserved as a testament to human survival in the process of natural selection. The human way of living, struggling, and fighting for its own survival must triumph. The truth discourse must validate its own claimed truth through the bodies and materiality of the giant panda.

The consequence of saving the giant panda from extinction is, understandably, the creation of the artificial giant panda. By "artificial", I do not mean breeding programs directly targeting the giant panda's fertility, but the creation of the giant panda as a species modeled after the human image. This recreated species exists in zoos and breeding centers, where giant pandas have been housed. In his postscript to *The Way of the Panda*, Nicholls (2012) suggests the emergence of the "virtual panda", which has little or nothing to do with the giant pandas roaming in the wild. It seems to me that this virtual panda has established itself as a distinct species from its wild counterparts through the structure of biopower and its roots in the visuality of the world picture. In this virtuality, the giant panda coexists with the human species and becomes fully dependent on humans for its survival.

The creation of the giant panda as a visual species ensures that the animal remains a means of production, super-exploited within capitalist relations. Saving the giant panda because of its cuteness is also a declaration of the achievement of capitalism in its most advanced form. On the one hand, new exchange relations are modeled on the abstract stereotypes of human baby cuteness, and profit is made available through the possibility of mobilizing the human species as a whole. In this consumption of the giant panda as a visual species – created anew and distinctively – the human species confirms the superiority of human excellence as a species. Becoming a new species does not exempt the giant panda from super-exploitation. Rather, cuteness captures the giant panda and regulates the species as a whole, meaning that the giant panda contributes to capitalist relations as a biopowered species. Here, we must consider how the capitalization of an individual giant panda differs from the transformation of the giant panda as a species. As Foucault (1990) notes, their integration into capitalism can involve different applications of biopower.

Foucault (1990) suggests that biopower functions on both disciplinary and regulatory levels. Disciplinary power targets individuals and centers on "the body as a machine" (Foucault, 1990, p. 139). It aims to increase the usefulness and docility of the body, integrating it into systems of efficient and economic control. The second, regulatory power, operates on a macro level, overseeing the "species body" and imbuing the species with "the mechanics of life and serving as the basis of the biological processes: propagation, births and mortality, the level of health, life expectancy and longevity, with all the conditions that can cause these to vary" (Foucault, 1990, p. 139). At its disciplinary level, Foucault's understanding of biopower overlaps with Marx's critique of the exploitation of the labor force – in that the working class is exploited by capitalists for the generation of surplus value, and disciplined individuals, no doubt, optimize the production of value. According to Foucault, social institutions such as schools, prisons, hospitals, asylums, and armies are deployed to shape individuals through disciplines. Suffice it to say that animal care institutions, such as zoos and breeding centers, are similar social institutions aiming to execute disciplinary power over nonhuman beings for the sake of generating more products and values – to the extent that disciplinary power continues to shape animals into more useful and docile individual bodies (see also Youatt, 2008).

Yet, the second level of biopower, namely, the regulation of population as species, binds nonhuman beings ever more closely to human society. Youatt (2008) suggests that the development of the regulatory form of biopower over nonhuman animals has been a response to the difficulty in disciplining nonhuman resources, such as justifying human needs and capitalist exploitation of nature as raw resources. Treating and regulating the nonhuman realm as a "new form of intervention in life itself" (Youatt, 2008, p. 401) take precedence in governing and managing nonhuman beings as self-disciplined subjects. It is with regulatory power that we must reconsider the role of zoos and animal breeding centers. Unlike disciplinary institutions where individuals are created, zoos and breeding centers are places that forge and shape visitors' awareness as a species. More specifically, they center the human species in governing the life forms, evolution, and populations of other nonhuman species.

Life and its forms are conceived in contrasting ways under sovereign power and regulatory biopower. While sovereign authority can suspend life through killing, regulation has the imperative to promote life to "optimise forces, aptitudes, and life in general" because its "highest function was . . . to invest life through and through" (Foucault, 1990, p. 139). This investment in life, however, does not fundamentally alter humanity's exploitative relationship with nonhuman participants. Instead, regulatory biopower continues the exploitation of nonhuman beings as a means of production, albeit through a different vein. Under sovereign power, humans held power over the lives of nonhuman beings – the killing of animals provided food, clothing, and other raw materials for various use values. Under regulatory power, nonhuman beings and their lives are promoted and invested in to the extent that near-extinct species, such as the giant panda, are redeemed, conserved, and preserved. Notably, in this life promotion and investment, the giant panda as a species remains a means of production, organizing the labor of an array of societal stakeholders.

Lastly, cuteness as a form of biopower shapes individuals through its promotion of self-education, self-discipline, and self-regulation, applying both to "the life of the body and the life of the species" (Foucault, 1990, p. 146). In other words, cuteness can discipline human bodies into individuals and regulate the human population as a whole species. As a disciplinary power, cuteness collaborates with social institutions such as schools and prisons to shape and form individual bodies according to expected social norms. This discipline attends to the smallest details of individual existences – it is not uncommon to spot visitors at Panda Base decorated with giant panda-themed accessories and altering their voices to resemble those of infants. As a regulatory power, cuteness functions as "an index of a society's strength, revealing both its political energy and its biological vigor" (Foucault, 1990, p. 146).

As the giant panda rises to become the national treasure of the Chinese, and the Chinese government deploys what has been termed the "panda diplomacy" (Hartig, 2013; Huang & Wang, 2019; Okafor et al., 2021; Xing, 2010) in its foreign politics, we witness how cuteness collaborates seamlessly with ongoing political and economic agendas as a means to regulate on national and species levels. The rapid growth of the giant panda fandom exemplifies the expansion of the disciplinary and regulatory reach of cuteness. The exploration of this point is the focus of Chapters 4–6.

Conclusion

This chapter begins with the question of how cuteness, as an aesthetic category for commodities, possesses the power to save a species on the brink of extinction. By examining cuteness and its roots in biological science, I delve into its connection to Foucault's concept of biopower. Here, it becomes evident that cuteness, as a form of biopower, indeed holds the potential to save – or even create – a species. It is also worth considering that the giant panda is not merely cute because she represents a cost-saving animal, as exemplified by Sir Scott's case, but rather because the giant panda serves as an invented species within the truth discourse provided by cuteness.

Revisiting Marx's concept of exchange value and means of production, I argue that the transformation of the giant panda into a cute species does not signify the animal's liberation from her super-exploited status. On the contrary, the investment in, conservation of, and promotion of the life of the giant panda species have been simultaneously integral to the establishment, improvement, and discipline of the human species. In Mandarin, the term for cute is **"可爱 (kě'ài)"**, which literally translates to "may be loved" – something that can be adored and appreciated. This Chinese conceptualization of cuteness urges us to examine the idea of cuteness as a form of biopower more closely. After all, what or who can be deemed worthy of love and adoration? The notion of being unworthy of consideration, love, and appreciation remains one of the most profound fears of the human species. For this reason alone, we may love the giant panda.

4

BECOMING PANDAS

Trade, diplomacy, and surplus

It is said that when Chinese and Western media present differing perspectives or cover certain issues in contrasting ways, Chinese media often turn to the giant panda to enhance their appeal to Western audiences. The specific strategy employed by Chinese media is to publish a giant panda video or image that quickly gains Likes from Western readers. The term "panda diplomacy" captures the ongoing dynamics between Western and Chinese interactions surrounding the giant panda. The use of the giant panda by Chinese media serves as a prime example of leveraging the animal's cuteness – in other words, using the giant panda as a political and cultural diplomat. While the complexity of panda diplomacy certainly warrants further exploration, my use of the term is limited to how it illuminates Western–Chinese interactions and the development of the giant panda as an international attraction.

In Chapters 1 and 2, I focused on showing how both the Chinese and the giant panda, in the early stages of transforming the animal into a global attraction, were largely silenced, dismissed, and marginalized despite their indispensable contributions to the capitalization and biopoliticization of the animal. I demonstrated that Western scientific discovery, imperialist ambition, and technological advancement became key terms in defining the emergence of the giant panda, while the Chinese were forced to collaborate under a variety of conditions that remained largely unarticulated in existing discourses. In this light, my use of the notion panda diplomacy is a shift from the unarticulated and silenced history of the Chinese in Western discourse to their more strategic, yet highly questioned, politicization of the giant panda as a diplomatic figure for the nation. How did this transition occur? If the Chinese were once deemed indifferent to the giant panda's appeal before becoming more actively engaged in its creation as an attraction, how can we explain this tremendous transformation? Could this easily lead to a dichotomy, portraying the

DOI: 10.4324/9781003511229-5

Chinese as either the ignorant Oriental Other or the cunning Chinese state? Are the Chinese simply schizophrenic?

This chapter seeks to find coherence in this inconsistency. Beginning with Chapters 1 and 2, my approach revisits the involvement of the Chinese in the creation of the giant panda attraction in its early historical moments. This means viewing the Chinese not as outsiders but as active participants in the capitalist production process and in the formation of "Chineseness" in the giant panda attraction from the very beginning. It also means that giant panda diplomacy should not be seen merely as a diplomatic tool or a point of pride while the animal symbolizes the cultural and economic shifts of the Chinese national subject since the establishment of the People's Republic of China (PRC) (Songster, 2018). Rather, to make the story coherent, we need to place panda diplomacy alongside the overlooked contributions, displaced bodies, and silenced experiences that are at the core of the creation of the giant panda attraction. The aim is not to restore glory or seek redemption for past actions. Instead, I intend to show how the capitalist agenda of transforming the giant panda into a global attraction, along with its rapid deployment of the necessary societal apparatus, has not loosened its grip but has instead necessitated modifications and transformations to accommodate its deepening exploitations.

The diplomacy beginning

Panda diplomacy refers to the "political undertaking . . . used by the Chinese government to win hearts and minds in selected foreign countries" (Hartig, 2013, p. 49). Panda diplomacy, which contributes to the most famous example of what can be generally termed "animal diplomacy", has performed at least three distinct roles: 1) Pandas given as gifts to enhance the reputation and status of the nation; 2) pandas exchanged as diplomatic gestures; and 3) pandas loaned to foreign nations (Hartig, 2013). Buckingham et al. (2013) reviewed the Chinese government's deployment of panda diplomacy and suggested that the government initiated a new, third phase of panda diplomacy. The first phase, according to the researchers (Buckingham et al., 2013), took place during the Mao era (1960s and 1970s), when China strategically gifted pandas to build friendships. The second phase began with Deng Xiaoping's rise in 1978, characterized by the giant panda gift loans that embodied "the capitalist lease model based on financial transactions" (Buckingham et al., 2013). The third phase emerged after the 2008 Sichuan earthquake when panda loans symbolized China's desire to acquire valuable resources and technologies, deepening trade relationships with foreign nations. Songster (2018) concurs with Buckingham et al. (2013), suggesting that the Chinese media began using the term "panda diplomacy" in the early 1970s when the Chinese government offered pairs of giant pandas to express goodwill and friendship to the people of recipient nations. In contrast, Nicholls (2012) traced the exchange of giant pandas as gifts back to the Nationalist Republic of China. In 1941, Chiang Kai-shek and Madam Soong Mei-ling offered a pair of giant pandas to the US government as a token of appreciation for US aid to the nationalists during the war with Japan. In her radio announcement covering the panda gift, Madam Soong expressed gratitude to United China Relief, the organization founded in New York City to raise funds for China during its national crisis.

In more recent studies on panda history, the Chinese historian Hong Jiang (2022) offered a more complicated view of these early stages of panda diplomacy. Commenting on the Nationalist government's gifting of the giant panda, Jiang (2022) considers the giant panda diplomacy a political negotiation involving a range of stakeholders, the most important of which included the US government, the Chinese government, United China Relief, and the Bronx Zoo. Once the giant panda established itself as a unique and remarkable attraction, setting new records for daily visits in various zoos, maintaining a constant display of giant pandas became a priority for European and US zoos. One major obstacle Western zoos encountered at this time was the Nationalist government's new policies on giant panda protection in 1939. According to Jiang (2022), the hunting of giant pandas since Ruth Harkness's success in 1937 had driven more Westerners to Sichuan and Muping, where giant pandas, both specimens and live animals, were hunted and sold at high prices. In a letter from the Sichuan provincial government to the Ministry of Foreign Affairs (cited in Jiang, 2022, p. 212), the Sichuan local government, concerned about the giant panda's survival, urged the Nationalist government to "stop foreigners from sneaking into the area and employing someone to go hunting in the mountains by paying high fees". The Nationalist government responded by taking a series of actions, including notifying foreign embassies in China and enforcing strict regulations on rare animal exports. The Chinese legal protection of the giant panda soon created problems for Western zoos, whose pandas often died of disease. To meet the growing market demand, zoos turned to alternative means of acquiring pandas. Jiang (2022, p. 214) notes that while Western media and researchers credited the Chinese government with the idea of gifting pandas, archival evidence from the Nationalist Publicity Department reveals that United China Relief initially proposed the idea.

While validating the details of these historical events certainly warrants a different project, what demands our attention here is how the same concept of "panda diplomacy" can mean contrasting stories for different historians. Let me add one more detail that will help make the story more coherent. Ruth Harkness, on her second journey to Western China to obtain another panda to pair with Su-Lin at the Brookfield Zoo, named her trophy after Madam Chiang Kai-shek as a gesture of honor (Croke, 2009). However, Harkness soon learned that "the mercurial Madame Chiang Kai-shek had gotten wind of Harkness's naming a panda after her and was feeling miffed instead of honored" (Croke, 2009). Harkness, within the day, rechristened the panda and later shared the incident in a private letter to her friend Anne Pierce:

> I hadn't been here a week, before a rumor came to me that Madam was much offended . . . A missionary here whom she (Madam Chiang) knows received a

> long letter from her secretary on the subject, and asked him to communicate with me, which he did. I swear, Anne, I didn't know whether to laugh or cry or have hysterics.
>
> *(Croke, 2009)*

Harkness's frustration over Madam Chiang's (over)reaction is not unique to the American designer in 1937 but continues to dominate the affective response when Western writers grapple with the Chinese's complex relationship with the giant panda. As long as the giant panda has been honored as the symbol of the nation, the pride of its people, and a political tool wielded by the Chinese government (Hartig, 2013; Songster, 2018), as the concept of panda diplomacy suggests, it becomes challenging to understand why and how Madam Chiang was offended. For Harkness and her biographer, Croke, the explanation for Madam Chiang's reaction was that she was simply "mercurial" or "hysterical". However, as I will explore in the following, Madam Chiang's reaction to the giant panda – specifically, her offense at the panda being named after her – may provide the only consistency with the responses of local hunters, landlords, and even the writer Chiang Yee, who lamented the naming of the panda "Ming" at London Zoo. This brings us back to the ongoing commercialization of the giant panda as a global attraction.

Diplomacy and trade

The panda diplomacy in 1941, as retold by Jiang (2022), suggests that the gifting of the giant panda was a political negotiation between the Nationalist Chinese government and several US organizations. If we do not rush to reduce Jiang's retelling to a national patriotic response in light of the fact that the Chinese historian was keen to demonstrate the political "correctness" of the Nationalist government, it is not too difficult to discern that Jiang's version of panda diplomacy helps build the consistency largely missing in Western scholars' works (Buckingham et al., 2013; Hartig, 2013; Nicholls, 2012; Songster, 2018). Rather than taking the leap to say that the Chinese transformed from being indifferent to the giant panda to a government skillfully manipulating the cuteness of the giant panda, it is perhaps more fruitful to see how the Chinese have collaborated – albeit not without reluctance, defiance, and struggle – with the transformation of the giant panda into an attraction all along. Panda diplomacy was, hence, first and foremost, an enforced response by the Nationalist government in order to meet the Western audiences' demand for the giant panda attraction. The tremendous market demand for the giant panda attraction and the capitalist drive for profit struck the key chord when Western zoos sought to replenish the animal. Gifting the giant panda, in this light, was a political undertaking made to sustain the global capitalization of the giant panda attraction.

Albeit contextualized in a different background, in *The Protestant Ethnic and the Spirit of Capitalism*, Chow's (2002, p. 19) analysis is illuminating in understanding the Chinese–Western political negotiation targeting the giant panda. Chow (2002, p. 20) made an insightful observation that is worth quoting at length:

> Accordingly, whereas the West asserts its moral claims on the basis of a universalist rhetoric traceable to the European Enlightenment, China is reduced to a reactive position from which it must and can speak only in terms of its own cultural and local specifics, in terms of its own historical differences . . . Instead of viewing one party, the West, as the holder of some absolute, uncompromisable value and the other, China, as a stubborn, wayward resister to all that is reasonable, it would perhaps be more productive, in light of Foucault's notion of biopower, to view the West and China as collaborative partners in an ongoing series of biopolitical transactions in global late capitalism . . .

Chow (2002, p. 21) discusses the collaboration between the Chinese and Western governments within the context of transnational corporatism, suggesting that political exchanges took place between the two parties to facilitate trade privileges and soften political rhetoric. Chow's argument highlights that, historically, individuals have sometimes been treated as commodities in international transactions, with both nations gaining benefits from these exchanges. Rather than focusing on the power dynamics between the West and China, Chow emphasizes how late capitalism has shaped both countries' collaboration in the exchange of biopolitical commodities.

The giant panda diplomacy that took shape in the 1940s, I argue, should also be understood through the lens of transnational corporatism, which sustained the global provision, production, and circulation of the giant panda as an attraction. By substituting Chow's concept of politicized individuals with the giant panda, and considering both China and the West as key players in the capitalization of the panda attraction, a more coherent narrative of panda diplomacy emerges.

When Armand David, the Roosevelts, and Ruth Harkness ventured to Sichuan to acquire the giant panda, the capitalization of the species was facilitated through Western institutions' commercial transactions with Chinese locals. For instance, David obtained the panda skin from a certain Mr. Li and shipped the specimen back to the French National Museum. The Roosevelts' hunting expedition was supported by the Field Museum of Chicago, with the brothers directly interacting with local hunters and landlords. Harkness, funded by the Brookfield Zoo, embarked on a second journey to China in search of another giant panda for the institution. Local assistance in her search was initiated through her own efforts.

These early, often discrete and disorganized exchanges of the giant panda between Western institutions and Chinese locals initially appeared to benefit the Western institutions and capitalists most. However, to expand the scale of capitalization, this institution-local-centered exchange pattern required urgent upgrading. The Chinese had two options: One was to become colonized and imperialized, thereby automatically installing capitalist institutions and mechanisms such as the zoo attractions; the other was to emerge as a nation-state, which could potentially offer flexibility and independence. History shows that China chose the latter.

The Nationalist government's gifting of the giant panda to United China Relief, where panda diplomacy took root, marked the next phase in the capitalization of the giant panda. It was no longer a one-sided domination of powerful Western institutions over powerless Chinese individuals. Instead, China was now able to respond as a collective whole, represented by the Nationalist government. Despite the capitalistic drive to further expand at this state-to-state level of exchange, the Chinese state power remained at an early stage. This meant that Nationalist governance was unstable, sometimes inefficient, and unable to sustain long-term exchanges. The imbalance was evident when the Western counterparts – comprising developed institutions such as zoos, private organizations, government bodies, and a thriving market – were contrasted with China's incipient state apparatus. Yet, the lesson for both the Chinese and Western nations was clear: The exchange of the giant panda, facilitated by nation-states, could elevate the transaction to the next level.

The West had to wait patiently until China was ready. It wasn't until Chairman Mao Zedong promised to send two giant pandas to an American zoo during President Nixon's 1972 visit to China that the giant panda began to catalyze the development of transnational exchanges. Buckingham et al. (2013) note that political strategies and friendly gestures were made through the image of the giant panda between China and the West at this phase. Simultaneously, conservation efforts initiated by the World Wildlife Fund (WWF) and the Chinese State Forestry Administration (Buckingham et al., 2013; Schaller, 1994; Songster, 2018) began at the Wolong Giant Panda Research Center. Collaborative research on panda biology also commenced there (Buckingham et al., 2013; Songster, 2018). While the giant panda exchange between China and the West during the Mao era was deeply intertwined with geopolitical agendas, this period did not exclude the capitalization of the giant panda attraction. Instead, the panda diplomacy, while serving political purposes, was firmly grounded in the market appeal and capitalizing potential of the giant panda.

The exchanges evolved into more complex and dynamic forms as both China and the West developed their state powers alongside market growth. In 1984, Deng Xiaoping made it clear that giant pandas were no longer gifts, but loans to be leased. This shift marked the establishment of the giant panda as a state-run industry. Becoming an industry meant that the market appeal of the giant panda took precedence over the political gestures and purposes that originally guided the exchange. In 1984, the Los Angeles Zoo received two giant pandas during the Olympics for a fee of US$ 5,000 per month per panda (Buckingham et al., 2013). Over the next 4 years, US and European zoos competed for panda exhibitions at the same rate, with eight zoos successfully securing them (Imbriaco, 2006). However, the rent-a-panda program soon faced intense controversy as animal welfare concerns grew (Schaller, 1994). Conservationists objected, citing the Convention on International Trade in Endangered Species, which stipulated that rare animals like the giant panda should only be traded for scientific purposes and/or to enhance species propagation and survival (Buckingham et al., 2013, pp. 2–3). The Chinese government suspended short-term leases in 1991, favoring long-term loans. This arrangement, supported by both Chinese and Western conservationists, ensured the breeding of giant pandas, while the lease fees funded panda management plans (Buckingham et al., 2013).

As the giant panda industry matured, the Chinese, bolstered by more developed state apparatuses, entered further collaborations with Western states across various business ventures, trade agreements, and construction projects. At this stage, panda loans were no longer part of larger deals but were seen as a "seal" of approval and intent for long-term, prosperous working relationships (Buckingham et al., 2013, p. 4). Simultaneously, the giant panda became a powerful cultural and symbolic animal, particularly as China emerged as a major world power (Weiss & Wallace, 2021) and strategically leveraged the panda's "soft power" (Gupta, 2022; Huang & Wang, 2020). The WWF's giant panda logo became one of the most successful trademarks of the 20th century (Barua, 2020) and remains emblematic of global conservation efforts.

One critical question for both China and the West, as their transnational collaboration on the giant panda industry flourishes, is how to move forward. Buckingham et al. (2013) suggest that the significant differences between how Chinese and Western cultures frame the human–nature relationship will be key to developing a successful panda conservation strategy. However, this point must be expanded: Because China and the West have consistently collaborated on the panda industry, this transnational project, which includes wildlife conservation, should be examined in its historical materiality rather than reduced to simple cultural differences. The key question in examining this historical materiality is why the Chinese were initially deemed irrelevant and passive, treated as mere political and economic tools in panda diplomacy, despite their integral role in the global capitalization of the giant panda attraction throughout history.

The repressive hypothesis

The politics of ethnicization, Chow (2002) argues, provides the central mechanism for mobilizing the transnational exchange of political victims between Chinese and Western states. Chow (2002, p. 22) suggests that dissidents, as commodities bearing the distinctive trademark of being "victims of their own culture, of being Chinese", underscore how ethnicity becomes central to generating a "sense of otherness, a foreignness" distinct from mainstream and normative society. Chow (2002, p. 23) further demonstrates how Foucault's repressive hypothesis crystallizes the politics of ethnicization by transforming "a negative, prohibitive idea into a positive, expansive set of possibilities". Within the constructed gap between "us" and "them", the ethnicized, imbued with the acute awareness of being Chinese – a culture and people marked by difference that, for humanitarian reasons, should ostensibly be overlooked –help ennoble the rescuing deeds of "us", the Westerners. This commodified relationship of ethnicity, as Chow (2002, p. 23) explains, is "underwritten by the conviction that the other is being held captive within his or her own culture, whether dead or alive, and that such captivity necessitates protest and liberation". Ethnicization, in forming the repressed negativity and the other, leads to a plethora of positive possibilities and activities, such as the commercial transactions of ethnic bodies, protests, and liberation movements.

Foucault's conceptualization of the repressive hypothesis emerges from his study of the history of sexuality (1990). He argues that, starting in the 17th century, sexuality began to be represented in a more repressive and restrictive manner, linked to peripheral deviants such as prostitutes, pimps, and perverts. Foucault writes that while modern societies may appear more prudish and restrictive about sexuality, this appearance is actually accompanied by a proliferation and intensification of discourses on sex. Modern institutions like medicine, psychiatry, prisons, and societal norms function as complex mechanisms that contextualize and provide a framework for the experience of sexuality. Moreover, the idea that the Victorian era marked a significant period of sexual repression supports a narrative extending through psychoanalytic explorations of repression to the sexual liberation movements of the 1960s and 1970s, and beyond (Heyes, 2014). This narrative underscores enduring societal discomfort with sexuality, urging individuals to confront silences and shame as a pathway to liberation (Heyes, 2014).

Foucault's study of sexuality and Chow's analysis of ethnicity highlight the two-step process upon which the repressive hypothesis is built. First, silenced sexuality and the othered Chinese are constructed as repressed entities within a restrictive economy through the politics of language and speech, which, in turn, drives broader social redistributions of sex and ethnicity. Second, and most importantly, the formation of this repressive hypothesis creates conditions in which the repressed must be liberated and emancipated. In this framework, repression and emancipation are "two tenacious collective beliefs" (Chow, 1996, p. 207) that operate in tandem. As Chow (1996, pp. 208–209) explains, the subject shaped and disciplined by the repressive hypothesis reaches its fulfillment through the pursuit of its own liberation.

> Whenever institutions mobilize themselves this way, we are in the realm, as Foucault's arguments show, not of personal liberation but rather of disciplinary subject formation. Would it be far-fetched to say that precisely those apparatuses that have been instituted for us to gain access to our "ethnicities" are at the same time accomplishing the goal of an ever-refined, ever-perfected, and ever-expanding system of visuality and visibility, of observation and surveillance, that is not unlike Bentham's Panopticon, the model of the ideal prison which Foucault uses to summarize the modern production of knowledge—a production that is inseparable from discipline and punishment, from the ubiquitous "economy" of sociopolitical control?

Emerged amid the repressed is "a veritable explosion" (Keat, 2013, p. 25) of discourses on ethnicity and sexuality, which perfects the modern subject through its visual representation and visibility. Let us now examine the giant panda industry and its prosperity within the framework offered by the repressive hypothesis. The initial construction of Chinese ethnicity, under the power of colonialism and imperialism, depicted local hunters and landlords as silent and passive contributors to the giant panda attraction, their efforts readily dismissed. By performing this role of the silent and passive other, locals were disciplined to the extent that their labor was capitalized and alienated as the giant panda was exported to Western nations. With the inclusion of state power during the era of panda diplomacy, the ethnicization of the Chinese as the other deepened, laying the groundwork for exploiting their labor. Within this repressed framework, the Chinese were assigned the significant role of embodying otherness. Their cultural, economic, and political differences increasingly became integral to the meaning and experience of being Chinese. Notably, the giant panda, like her Chinese counterparts, became an ethnicized animal and embodied this process of ethnicization.

The 1931 exhibition of giant pandas, hunted by the Roosevelts and displayed at the Field Museum of Chicago, presented them in "their favorite habitat of bamboo thickets which in western China are found growing at altitudes up to 10,000 feet or more" (Osgood, 1931, p. 1). The pandas' Chinese background was framed through natural landscapes and scientific facts. Ming, the giant panda at the London Zoo, gained immense popularity and "played an important role in boosting the morale of wartime Londoners, who flocked to visit the zoo's star attraction" (Hussey, n.d.). Chiang Yee, while writing about Ming, considered the panda a true representative of China, attributing to her virtues such as good nature, friendliness, patience, and persistence, which he associated with the Chinese. When Pan-da and Pan-dee, gifts from the Nationalist government, arrived in New York on December 30, 1941, amid the Pearl Harbor attack, they were considered a special Christmas gift from the Chinese to the Americans. This early stage of panda transactions, driven by Western markets and led by Western states, maintained friendly terms with China. Like the giant panda, the Chinese were perceived as meek, benign, and in need of help, while the transactions of giant pandas consistently demonstrated the West's humanitarian and kind gestures.

However, these friendly terms, which linked the Chinese and the West through the gifting of giant pandas, were abruptly suspended after the Communist government consolidated power in China in 1949. Chi-Chi, a panda from this period, was labeled a "communist good" and barred from entering the United States, ultimately finding her way to London. According to Nicholls (2012), this rejection stemmed from Chairman Mao's probing question in August 1949, "Why can't we live without the United States?" and his revolutionary stance toward foreign powers. Songster (2018) supports Nicholls, noting that the Mao's leadership in the late 1960s marked the rise of the giant panda as a national icon. She concludes that the political power of the giant panda lay in "its innate ability to exude an apolitical image".

In the same period, the giant panda came to symbolize the "extreme nationalism and self-reliance" of the Chinese (Songster, 2018).

This dual force – the United States' marginalization of the giant panda as a communist good and China's response to build a self-reliant nation – can be seen as two sides of the same coin. Through the lens of the repressive hypothesis, the Chinese were both forced and compelled to transform into "the Chinese". On the one hand, Western nations identified the Chinese as representatives of a distinct political regime and as another pursuing divergent political agendas. This Chineseness, embodied by those within China's geographical borders, became associated with cultural differences, political deviance, national excessiveness, freedom deprivation, and distinctive movements. On the other hand, China's self-reliance, ostensibly a nationalist reaction against Western marginalization, also reflected the necessity of creating a complete other to mirror Western counterparts. In this framework, the emergence of a self-reliant China was not merely a display of emotional fervor but a mandate for achieving a cohesive and unique national identity. Ultimately, the global giant panda industry was a result of this transformation, as transnational corporatism increasingly demands stately engagement and biopolitical organization.

When the term "panda diplomacy" became widely employed in Western discourse to describe China's use of the giant panda's cuddly appeal to win the hearts and minds of selected Western nations (Hartig, 2013), the Chinese were framed as an other whose political undertakings underscored their differences. Marked by their political distinctiveness, the Chinese were portrayed as indifferent and cunning, capable of leveraging even the world's most endearing species as a mere political tool to serve their national interests. As Songster (2018) observed, "state-gift pandas were among the most successful efforts by China to show a new face as it strove for greater international recognition and integration during the early 1970s". Songster (2018) further noted that the giant panda allowed the world to foster friendly relations with the "inscrutable Communist power", while earlier exchanges of economic and strategic partnerships gave way to the nation's identification and embodiment with the giant panda. As "panda diplomacy" framed China's state power within a distinctively Chinese approach to politics, it also tethered the Chinese to "its own cultural and local specifics, in terms of its own historical (political) differences" (Chow, 2002, p. 20). The giant panda diplomacy asserted its moral claims through a universalist discourse that could arguably be interpreted as reinforcing the West's perceived superiority in political, ethnic, and cultural terms.

What does it signify that the giant panda becomes corporeal to the Chinese, enabling them to speak both through and for the animal? How do we interpret Madam Chiang's burst of anger upon learning that Ruth Harkness had named a giant panda after her? If Madam Chiang had foreseen the giant panda's transformation into a national icon, symbol, and source of pride, would she still have rejected Harkness's kind gesture? These questions lead us to consider what it means for a nation to fully identify with the giant panda – a new face through which China engages in international affairs and exchanges – and what this deep association ultimately entails.

Becoming the panda

An early morning of my workday at Panda Base typically began with rising at 6:30 and ensuring I arrived at the office on time. However, a text message from a former friend disrupted my schedule that day, as I took time to reply to her. In her message, she shared that she had been crying for a week over the mistreatment panda Ya Ya allegedly endured at Memphis Zoo. She sought answers, believing that, as someone closely affiliated with pandas, I could explain why such suffering occurred. She expressed sympathy for Ya Ya and her hope that the panda would return home soon. I found it challenging to respond to her emotions, especially her confession of crying over the panda. Why this excess? Yes, Ya Ya was a national treasure, a precious panda, but animals can fall ill, and they do not remain healthy forever. Moreover, the intense dynamics between Chinese and Western media coverage often meant one could not take every claim at face value. Although I was vaguely aware of Ya Ya's story due to her ubiquitous presence on social media in early 2023, I felt impatient with my friend's lamentations. Why couldn't she – an accomplished collaborator in English–Chinese translation projects and someone with a master's degree in English literature – recognize that Ya Ya's story was likely media hype? Perhaps a social media buzz story? I decided that despite my affiliation with a panda organization, I was not responsible for indulging her emotional engagement with a viral narrative.

The case of Ya Ya, however, had drawn widespread attention. Chinese media and panda enthusiasts expressed emotional outbursts regarding her treatment (Magramo, 2023; May & Hauser, 2023; Wang, 2023). Upon her return to China on April 27, 2023, millions welcomed Ya Ya at Beijing Zoo, sparking what Magramo (2023) described as a "fresh outpouring of online pride". Cheng (2023) noted that Ya Ya's return became "the subject of online nationalistic fervor", culminating in a campaign that mirrored the growing tensions between Washington and Beijing. Interviewed Chinese students and activists declared that the giant panda would "definitely be better off in China" (Cheng, 2023). While the Memphis Zoo, in collaboration with Chinese experts, maintained that Ya Ya had not been mistreated during her 20-year stay in the United States, this explanation was not widely accepted in Chinese online communities (Magramo, 2023). The Memphis Zoo even created a page on its website titled "Memphis Loves Pandas" to affirm its care for the pandas. According to a follow-up report by In Defense of Animals (Michelson, 2023), Ya Ya now resides in an enclosure inaccessible to tourists, enjoying a peaceful retirement.

The public outcry over Ya Ya's perceived mistreatment reflects a significant shift: The Chinese public has transformed from passive participants in the early days of the giant panda industry into outspoken contributors to the transnational panda narrative. In light of China's rising power and growing emphasis on soft diplomacy, the activism surrounding Ya Ya sharply contrasts with the earlier passive myth-making and children's storytelling that defined Chinese engagement with pandas. Unlike the unnamed panda killed by the

Roosevelt brothers or Ming, christened by the London Zoo, Ya Ya became a focal point of protest and resistance. This protest against perceived mistreatment at Memphis Zoo and the online declarations celebrating Ya Ya's return to her homeland embody the subjects of "Chineseness" under the framework of the repressive hypothesis.

Historically, the body of the panda has been a site of protest and meaning-making. In Chapter 1, I discussed how the panda exported to Chicago by the Roosevelt brothers was found dead before her exhibition debut. Her death became a site for emotions unique to the Chinese community and a form of protest against the commodification of pandas and the exploitation of Chinese locals. Moving forward, Chiang Yee's story of Ming reveals a different dynamic. Ming, with her perpetually sad face, remained in London Zoo, aware that her old home was a place she could never return to. Ming's Chinese subjectivity was characterized by homelessness – a detachment from both her origin and her present surroundings. Chiang idealized Ming's patience, good nature, and temperance, framing her as inherently "Chinese".

With Ya Ya, however, "Chineseness" has taken on a more material and explicit form, centered on the concept of a home country and nation. The protests from my friend and the Chinese online community articulated a belief in a home that welcomed the mistreated panda. The homelessness experienced by Ming and her predecessors has been inverted: Ya Ya, unlike them, has a place to live where she would neither be mistreated nor exploited. Ya Ya's farewell party in Tennessee, attended by about 500 people and featuring Chinese cultural performances and goodbye letters, signaled this shift. During the event, Ya Ya was "surrounded by bamboo and given a special ice cake made of grapes, sugar cane, and cookies" (Wang, 2023). Millions of netizens commented online, celebrating her return to China. The abundance of cultural symbols – food, Internet comments, performances, and letters – used to identify Ya Ya as "Chinese" stands in stark contrast to the deprivation endured by earlier pandas. This abundance underscores a new phase in the representation and experience of Chineseness, with Ya Ya embodying a shift from sacrifice to affirmation.

Interestingly, the Chinese public's sentimental outburst and excessiveness about Ya Ya align with a gradual unraveling of panda diplomacy. As Anderlini (2017) observes:

> Many zoos and governments remain keen to make a deal with China to receive these cuddly emissaries but the proliferation of politically charged panda loans has already lowered their value. Faced with a tortuous diplomatic negotiation process, many Beijing-based diplomats now dread the prospect of engaging in panda diplomacy. "I really hope we don't do it," says one European diplomat from a country that does not have pandas but may officially request them soon. "It's just not worth the expense or the hassle."
>
> This illustrates the central problem of panda diplomacy today. The process has become onerous and expensive, the conservation benefits are dubious and

> there are now so many pandas outside China that more and more countries are likely to decide they would rather not bother. If that does happen, then not only will modern panda diplomacy have undermined, rather than enhanced, China's soft power, it will have damaged this iconic animal's chances of survival in the wild for nothing.

The growing complexities of maintaining panda diplomacy stem not only from its high costs but also from its increasing entanglement with political agendas, which now act as significant obstacles to further development of the panda industry. The repressive hypothesis – that China represents a different political regime as a collective other – begins to falter as the effort to sustain this negation grows riskier and more challenging due to China's expanding global influence. Ya Ya's story, marked by the temporal political tensions between the Chinese and US governments, reflects the collapse of framing China solely as a political and stately other. This negation and repression must now be relocated into different frameworks and bodies through which the giant panda industry can continue to harness the forces of capitalism.

The protesting Chinese public, having emerged as disciplined individuals, forms a fertile ground for the further capitalization of the giant panda industry. Okafor et al. (2021) examined whether panda diplomacy promotes Chinese outbound tourism and found that Chinese tourists are more likely to visit countries hosting giant pandas. Their research suggests that post-pandemic panda-themed international marketing campaigns could effectively attract Chinese visitors. Similarly, Guo and Fennell (2023) found that visitors tended to learn, donate, and engage more in the presence of giant pandas compared to other species at the Panda Base. These studies highlight the implicit and embodied connections between Chinese and panda-hosting destinations, as well as their interest in further investing in the giant panda. From this perspective, Ya Ya's farewell party can be seen as a triumph of these everyday economic arrangements: Commodities and experiences linked to Ya Ya – including bamboo shoots, cultural performances, letters, an idealized notion of "home", and online posts – offer avenues for Chinese individuals to express their identity and align with the giant panda.

Foucault's approach to discipline begins at the microlevel, where local relations constructed through individuals' behaviors, interactions, and modes of expression are orchestrated into larger patterns such as national norms or regulations (Lynch, 2014). As individual bodies and individuality are products of disciplinary power, these individuals also evolve into useful and obedient patterns that are well-regulated and controlled. This production of individuals and individuality is a consequence of the ascendancy of life over death. The centrality of making life in a disciplined society, coupled with the force of capitalism, measures the existence of individual beings through their integration into the process of capitalization. Ethnicity, or more specifically Chineseness in this study, should be understood "as part of an already biopoliticized economic relation, whereby the very humanity attributed to ethnics is itself firmly subsumed under the process of commodification and its asymmetrical distribution of power rather than outside them" (Chow, 2002, p. 32). Ethnicity, for Chow, is no different from Foucault's concept of sexuality, whose reproduction begins with an array of apparatuses and institutions intersecting and disciplining the human body.

It is the giant panda industry's increasing power to make life – and, in this case, to create a Chinese identity – that, I believe, is the driving force behind what I term the giant panda fandom among the Chinese community. The fact that the life of the Chinese could be lived so intimately and emotionally through one's identification with the giant panda by means of commodity exchange suggests that the giant panda industry may grow at the expense of individuals' engagement and participation. It suggests that the industry has shifted its focus from state power, traditionally represented by panda diplomacy, to the currently disciplined Chinese individuals through its meticulous commodification and capitalization processes. The consumption of giant panda-related discourses and products, readily available to individuals, provides access to a population of 1.4 billion – the biopower and its control of a population at its zenith.

As a biopoliticized economic relation, the ethnic life of being Chinese completes itself not only through the consumption of commodities but also through protest and resistance. Ya Ya's homecoming campaign served as the ideological force allowing the Chinese to imagine a homeland where the animal, along with their Chineseness, could be ultimately saved, preserved, and redeemed. This idealization of a home place – a local specificity and geographical boundary safeguarding the animal – materializes the meaning of Chineseness. The giant panda was no longer a homeless animal trapped in foreign zoos but an animal with origins and materiality. This marked the transition of Western–Chinese differences – more specifically, the construction of the Chinese other – from cultural forms such as political states and cultural symbols to what appears to be natural and environmental: The landscape and geographical boundary itself. The fact that a region on this planet, outlined as China, does not cross over or intersect with other cultures and Western nations provides solid scientific knowledge – at least within the time and space of this Earth – that delineates the boundaries of human experiences and limits. What might be termed the "geopolitics" of the giant panda industry, through the mobilizing and politicizing of the animal's image, offers a path to understanding the ongoing capitalization of the giant panda. In this framework, conservation and preservation, while important, are secondary roles compared to the bioeconomics that the second-largest population on the planet – the Chinese – could generate on a global scale.

The surplus: Chinese and cute

Although the repressive hypothesis allows us to see how oppression, protest, and resistance are integral to the Chinese–Western exchanges of the giant panda through local, stately, and commodity forms, a question that continues to trouble me is why and how China's consistent participation in the capitalization of the giant panda has been disregarded and undermined. While Chow's (2002) study of the exchange of political individuals between China and the West exposes both as participants in an ongoing transnational corporatism, I believe it is also necessary to investigate how the denial and negation of China's engagement help affirm and consolidate this bio-transnational transaction. The giant panda industry offers an excellent example for this further investigation.

Chapters 1 and 2 show that the Chinese's initial participation in the giant panda attraction, mostly on local and individual levels, has been consistently dismissed and bypassed in existing Western discourses on the giant panda industry. For example, neither Nicholls (2012) nor Songster (2018) suggested that the Chinese played an active role in the giant panda's discovery and the early Western trophy hunting of the animal. In contrast, I demonstrate how local myth and Chinese writers made laborious efforts to make sense of these events both physically and mentally – the mythical death of the giant panda and the loss of a homeplace were among the prominent tributes paid to the dominance of imperialism and colonialism at this stage. Moving on to panda diplomacy, China's success in winning the hearts and minds of the Western public was considered as political strategy and manipulation because the Chinese were conceived as the political and stately other. Hence, Hartig (2013) refers to the giant panda as China's "most unique diplomatic tool", while the Western zoo and public's requests for the giant panda attraction were scarcely mentioned. The recent homecoming of Ya Ya, fueled by Chinese nationalist patriotism, was similarly framed as a consequence of intensified China–West interactions.

What is missing in this expanding global industry of the giant panda attraction, interestingly, is the consistent exchange of the giant panda and the engagement of both the Western and Chinese parties as industrial partners. In this global industry, the Chinese have assumed a series of roles – the passive, silent, and inapt colonized; the elusive and strategic manipulator of politics; and the emotional and passionate nationalists. However, their consistent participation in the giant panda industry, as partners in the West's capitalist projects, has yet to be acknowledged. If the giant panda industry, like the exchange of political individuals between the West and China, represents transnational corporatism, what does it mean that one partner in this collaboration has never been recognized for its corporate contributions but only through roles performed outside this corporate framework? Why, we must ask, is panda diplomacy not seen as a collaboration between two capitalist ventures that mutually understood the values their cooperation could generate, but instead reduced to one party's political cunningness? Why is the panda trophy hunting and discovery not framed as an exchange between locals and Western empires, but rather as the locals' indifference and passivity? Why is Ya Ya's return not celebrated as a triumph of capitalist consumerism and commodities but only perceived as the nation's patriotic protest?

To answer these questions, it is necessary to bring the repressive hypothesis back to its roots in biopower and biopolitics. The repressive hypothesis, in essence, provides a means to control and manage a population toward more productive roles and behaviors within the capitalist regime. This implies that when the Chinese are framed as the silent and passive other, the different political state, or the protesting nationalists – rather than participants in capitalist transactions – their productivity, or the experience of "making life", can be optimized rather than diminished. Regulated as "the Chinese", this life experience of a nation's population aligns with what Cooper (2011) refers to as "life as surplus", a resource that can be reaped, circulated, and regenerated without exhaustion. The Chinese surplus, relying on the 1.4 billion human population, has a value that the giant panda, with approximately 2,500 individuals alive despite being one of the most "expensive" animals, cannot equally match.

In her study of political shifts in postwar United States, Cooper (2013) demonstrates that fundamentalist views on gender, sex, and race were invoked not merely as cultural battles distracting from the real economic issues of capitalism but as integral to the functioning and perpetuation of capitalist ideologies. In the giant panda industry, defining the Chinese as a cultural, political, and nationalist other – rather than as an equal partner in capitalist exchange – shifts the responsibility for the maintenance and operation of capitalist processes onto the Chinese while significantly reducing the perceived impact of the giant panda's exploitation.

In his examination of the medically commodified body in contemporary Chinese and transnational culture and media, Heinrich (2018, p. 5) suggests that the rhetorical and visual traditions of a "universal" human undermine the creation of a distinctly "Chinese (or 'Chinese') human", whose body serves as "a source of profit whose humanity is qualified or conditioned by its availability as a kind of global corporeal surplus". Heinrich notes that promotional materials and microcultures surrounding the traveling plastinated cadaver exhibits perpetuate colonial racial dynamics by endorsing an implicit division between the "human" (first-world viewers with superior ethical standards) and the specimens (nonwhite or subaltern bodies stripped of ethical consideration and displayed). The Chinese surplus refers to how Chinese corporeality has been rendered a "renewable resource" made possible through "the intolerable abjection of its own origin" (Heinrich, 2018, p. 117). His study reveals that Chineseness has been mobilized as an "organic" or "biological" force within the bioeconomy, adding value to the expanding global art market through embodied corporeality.

How does this Chinese surplus, in the giant panda industry, intersect with the aesthetic category of the cute? If Chineseness is the biological source of profit that a nation's population affords through labor, cuteness, I argue, provides an aestheticized venue where this Chinese life interacts with the critical means of production of the giant panda industry: The giant pandas themselves. In her study of the gimmick, Ngai (2020, p. 17) connects this aesthetic category to issues of gender and race, asking: "Given the consistency with which gender and race get written off as peripheral to capitalism, does the relation of those marked by these categories to the superfluous/necessary gimmick become especially charged?" Substituting "gimmick" with "cuteness", this question encapsulates the particular relationships established between the Chinese and the giant pandas. Ngai (2020, p. 17) further shows that the gimmick is an adaptive and flexible aesthetic category that reflects "the different strategies and attitudes of those trying to navigate and survive within a capitalist system".

Studies have demonstrated that perceptions and uses of cuteness vary based on participants' social and economic positions, particularly in relation to race and labor. For example, in his study of kawaii culture in Japan and cuteness in the West, Dale (2023) shows that nurturing intimacy toward babies is interpreted differently depending on cultural contexts. He suggests that the group-oriented values of Japanese society emphasize the bonding force of cuteness to bridge child–adult and gender divides. In Japanese culture, *kawaii* thus symbolizes mutual dependence rather than individualism. Takashi Murakami links Japan's *kawaii* culture to its postwar context, suggesting that Japan embraced cuteness to appear harmless and inoffensive to its conqueror, the United States (Dale, 2023). Allison (2003), using the example of Pokémon capitalism, demonstrates that the aesthetic of cuteness integrates hyper-consumerism, technological advancement, and portability to promote the global Pokémon market. In contrast, Dale (2023) argues that in the United States, cuteness is associated with cleverness due to the nation's emphasis on self-reliance and individualism. In American culture, cuteness carries "a note of nonconformity or a hint of cunning rebellion" that moves beyond its infantile traits. The national character of the United States, therefore, reflects cuteness as a symbol of independence and ingenuity, diverging from its interpretations in Asian cultural contexts.

It is at this intersection of cuteness and Chineseness that I believe the phenomenon of panda fandom must be scrutinized within the societal and economic contexts in which it forms. Speciesism, accompanied by the aesthetic of cuteness, and ethnicity, expressed through Chineseness, are two biopolitical forces that have bonded the Chinese to the giant panda in myriad ways. On the one hand, there is the historical West–China interaction, where the Chinese have been consistently repressed and othered, making Chineseness synonymous with marginalization and ongoing protests about one's identity. The giant panda industry has created social exchanges and relationships through which Chinese subjectivity can be individually and privately constructed as people engage with the animal. On the other hand, cuteness operates through the power of making life, modeling the species' existence and survival on the image of the human infant. This dual process – of becoming a human species and becoming Chinese – ties the Chinese to the giant panda through both subjective identification and objectifying experiences of embodiment. This examination, however, remains incomplete without a deeper exploration of the roles the giant panda has played within the giant panda industry. The embedded corporeality between the Chinese and the giant panda suggests that understanding the panda's dilemmas and experiences within the industry can, in turn, illuminate how the Chinese contribute to the generation of surplus value in transnational exchanges. I argue that the giant panda's super-exploitation, as a means of production, has been one of the cornerstones supporting the industry's expansion.

Ngai (2020) describes the intense interaction between living labor – active, present-day work performed by workers during the production process – and dead labor, the means of production materialized through the embodied past labor, as central to the capitalist valorization process. The ratio of dead to living labor embodied in an artifact determines its value and the rate of profit. In this context, panda fandom represents a field where capitalist relations experiment with two of their essential participants: The Chinese public, as the main contributors of living labor, and the giant pandas, as the embodied means of production. These participants collaborate in the production process. Panda fandom, characterized by a deeply personal, individual, and loosely connected relationship between the giant panda and the Chinese, as the next two chapters will show, provides an open field of capitalization where human and nonhuman participants are continually experimented with and transformed into new bodies, experiences, and biological futures.

5

PANDA FANDOM

Confession, celebrities, and primitivity

In this chapter, I explore how panda fandom emerged as a biopoliticized population through the Chinese public's engagement with the contemporary forms of the giant panda industry. As panda diplomacy begins to lose its power to define and control the Chinese population as the political and cultural other, a new governmentality technique becomes urgently needed to manage and control the world's second-largest population. Central to this transition in governmentality is the disciplining and production of individuals whose corporeality and bodies become the building blocks of panda fandom. This individuality, materialized through the fan's personal and private labor, reflects a lived and embodied experience unique to panda fans. The fanship connecting fans to the giant panda, in this light, creates a "life-enabling" or "life-making" space that amplifies the giant panda's attractiveness. Panda fandom, I show, offers a space where fans' heterogeneous and uncoordinated acts of production can be freely and randomly registered, even though these personal contributions can only gain meaning – and thus become exchangeable – through abstract ideologies such as cuteness and Chineseness.

Panda fandom serves as a site where the biopolitical forces of cuteness and Chineseness are collectively labored by fans and integrated into capitalist relations, which, in turn, provide fans with the subjectivity and individuality necessary for their experiences in panda fandom. The fanship linking fans to the giant panda captures the specific relationship that fans, as individual laborers, establish with the giant pandas – the means of production within the giant panda industry. In this context, panda fandom represents a form of governmentality designed to exercise power over a population that is engaged with notions of cuteness and Chineseness.

DOI: 10.4324/9781003511229-6

Fandom: A biopolitical perspective

The term fandom was first investigated in cultural studies based on seminal works by scholars such as Stuart Hall. One of the foundational texts on fandom, Jenkins's (2012) *Textual Poachers: Television Fans and Participatory Culture* provides key concepts that have significantly influenced scholars' understanding of fan communities and their dynamics. Drawing on Michel de Certeau's notion of "poaching", Jenkins (2012) considered fans as "poachers" who actively participate in appropriating, reinterpreting, and repurposing media texts. As textual poachers rather than passive recipients of media, fans are creative, dynamic, and sometimes subversive participants in participatory culture. In Jenkins's (2012, p. 46) own words, fandom becomes a "participatory culture which transforms the experience of media consumption into the production of new texts, indeed of a new culture and a new community". Jenkins's (2006) follow-up work, *Convergence Culture*, expands on the notion of fandom as a participatory culture by incorporating newly emerged media industries. Key themes that have emerged following the exploration of participatory fan cultures include identity formation (Aden et al., 2011; Gray et al., 2017; Hills, 2003; Piacente, 2020) and affective attachment (Duffett, 2013; Lipsman et al., 2012; Sandvoss, 2005; Stanfill, 2019).

Jenkins's approach to fandom as a culture where communities engage to establish identity and negotiate meanings establishes the premise that cultures are seen as taken-for-granted givens. However, further studies on fandom have increasingly problematized fandom as cultures, aiming instead to conceptualize fandom as particular social constructs situated within specific empirical realities and historical materialities. Stanfill (2015), for instance, considers fandom as a discourse produced within a discursive environment whose structure is rooted and resistant to change while also channeling and orienting participants' actions and behaviors in specific ways. A significant theoretical shift to place fans and fandom at the center of cultural practices and media relations is achieved by investigating fans' roles within the capitalist structure. As Stanfill (2019, p. 4) concludes, fans are treated as both "subpublic" and "counterpublic" groups that build unique relationships with a universal and overarching notion of the public. Suggesting that fans are "acted on as a population through producing, disseminating, and reinforcing a norm of media use", and drawing on Foucault's notion of biopower, Stanfill (2019, p. 10) argues that fans are not managed, governed, and regulated within the industry as individuals or cultures but as a population whose life processes are optimized and oriented. Stanfill (2019) proposes that the domestication of fans serves as the organizing metaphor for understanding the biopoliticized fan public.

Fan labor, according to Stanfill (2019), as its distinctive discursive construction indicates, allows fans to perform work and labor that are not traditionally rendered or defined as labor. The exploitation of fan labor, Stanfill (2019, p. 19) writes, can only be understood in relation to "labor-cost reduction on industry's part, the rise of pleasurable labor, and the rejection of capitalist projects by many fans". Precisely because this particular form of fan labor is biologically embodied, unknowingly performed, and coerced at the population level, fans willingly consent to contributing their workforce to the industry. Milner (2009) introduces the notion of *New Organization* to studies on fan labor, showing that the concept, which emphasizes the specialized labor of self-motivated "knowledge workers", readily justifies the autonomous and collaborative nature of fan labor. Milner resonates with Stanfill's point that fans have accepted uncompensated labor as a premise and have developed devoted commitment to the texts they produce. Busse (2015) likens fan labor to reproductive labor traditionally performed by women, arguing that because capitalism has historically suggested that love and labor may be anti-proportional, fan labor, driven by love and passion, is continuously devalued. De Kosnik (2012) writes that fan production is a category of work that is badly needed in an era of fragmented markets.

A biopolitical perspective on fandom conceptualizes fans as a population whose biological and bodily aspects are regulated through power. According to Foucault (1990, p. 25), a population is an innovation in the techniques of power since the 18th century and is addressed as an "economic and political problem". Governments regulate populations through their specific phenomena and peculiar variables, such as birth and death rates, states of health, patterns of diet, and habitation. This power to control a population also defines what a government and its stately power do. Foucault (1990, p. 25) cited Herbert to support this observation:

> States are not populated in accordance with the natural progression of propagation, but by virtue of their industry, their products, and their different institutions . . . Men multiply like the yields from the ground and in proportion to the advantages and resources they find in their labors.

Fandom, understood through Foucault's notion, refers to a regulated population whose biological features are specifically managed to yield economic and political benefits for the nation-state. Hence, Foucault's concepts, such as governmentality, power, and discipline, are especially useful in understanding what fandom and being a fan entail in contemporary society. Stanfill (2019) has elaborated on fandom using these ideas. However, while Stanfill (2019) aims to combat the exploitation of fans by defending the concept of fans' free labor, I suggest that viewing fans as a population regulated by the state also implicates their management through industry, products, and other institutions. Fans' free labor is integral to the creation and maintenance of this population, while fans themselves are individuals disciplined to align with the values and visions of stately institutions.

I therefore diverge from Stanfill's (2019, p. 173) argument that the "mismatch between fan and industry values" is the primary cause of fan exploitation. Rather than viewing fans as individuals dichotomized from the industry, I argue it is more critical to understand fans and the industry as participants and collaborators in the regulation of fandom populations. This interdependent relationship between fans and the industry signifies not only that fans are individuals disciplined to provide free labor for the industry but also that the effort to "respect what this work (free labor) experientially means" (Stanfill, 2019, p. 173) is already embedded within the rationalized process of population economy itself.

Through Marx's terms, we can further examine the interdependent relationship between fans and the industry. In Marx's analysis, heterogeneous and uncoordinated acts of personal labor do not automatically translate into value. Instead, the abstraction of this personal labor – its integration into broader social divisions of labor and production – provides the means and measurement to understand personal labor as value created (Ngai, 2020). Fans, in this light, are individual actors whose personal engagement with fandom generates a plethora of activities and living labor performances. However, the evaluation of their labor depends on a retroactive exchange of their products through social mediations often facilitated by industrial organizations. Acknowledging and respecting the personal and living labor of individual fans can temporarily ease tensions between fans and the industry, but it does not resolve the gap between living and abstract labor that determines value creation and surplus valorization in capitalism. In fact, experiencing the abstraction of individual labor can be critical for fans, as this exchange process enables them to interact and form relationships with the world around them – a process vital to the survival of human individuals within any social construction (Ngai, 2020). Fans, as free workers, and the industry, as the entity abstracting and exchanging free labor, become interdependent participants in the capitalist mechanism. To better understand how fans, both as individuals and as a population, experience the regulation of stately power, we must turn to another concept from Foucault: Confession. This framework allows us to approach fans and fandom through their specificities, shedding light on the nuanced ways in which fandom is regulated and maintained.

Fans and confession

Confession is a central practice in contemporary Western societies, serving as a means through which individuals understand and govern themselves. As Foucault observes (1990, p. 59), "Western man has become a confessing animal". The technique of confession, characterized by a meticulous process of self-examination, is invoked when sexuality – the focus of Foucault's (1990, p. 44) investigation – becomes a medical and medicalizable object. Individuals are compelled to detect and analyze sexuality as "a lesion, a dysfunction, or a symptom". This process of detecting and examining sexuality as a symptom is extensive, effective, and meticulous, but it also entails a "sensualization of power and a gain of pleasure" (Foucault, 1990, pp. 44–45):

> This produced a twofold effect: an impetus was given to power through its very exercise; an emotion rewarded the overseeing control and carried it further; the

> intensity of the confession renewed the questioner's curiosity; the pleasure discovered fed back to the power that encircled it. But so many pressing questions singularized the pleasures felt by the one who had to reply. They were fixed by a gaze, isolated and animated by the attention they received. Power operated as a mechanism of attraction; it drew out those peculiarities over which it kept watch. Pleasure spread to the power that harried it; power anchored the pleasure it uncovered.

In this paragraph, Foucault describes a cyclical and reinforcing relationship between power and pleasure, particularly through the techniques of confession and surveillance. The exercise of power generates its own momentum and emotional rewards, while the intensity of confession stimulates further curiosity. This process singularizes and isolates the pleasures of the confessor, making them more pronounced and under the control of the overseer. This dynamic creates a feedback loop where power and pleasure continuously reinforce each other, ensuring the perpetuation and deepening of the power structure.

This "confessing animal" of Western man occupies Mitchell's (1991) outsider–insider position when interacting with the visual attraction, which contextualizes the individual within the visualized world around him. As mentioned in Chapter 2, this tourist figure, both an insider and an outsider, experiences the paradoxical detachment and close attention to the attraction. This very effect of having a position (Mitchell, 1991) creates the individuality and subjectivity that Foucault's (Hoffman, 2014) notion of power disciplines and controls. A viewer of the world-as-picture, the tourist exercises power over the image, a process that rewards the tourist with pleasure and emotional intensities. The tourist, as an individual, is responsible for experiencing this pleasure and emotion as the reward. This is the critical moment where confession comes into the picture. Having received the pleasure, the tourist is invited to examine this specific pleasure through details experienced and how the body, as a manifestation of this pleasure, encodes the recipient of this pleasure. This confession of the tourist is the technique that binds the tourist to the world picture as the attraction's insider. The confessional process ensures that this pleasure – the symptom treated and recovered – is experienced as isolated singularities, amid which the individual and individuality emerge. Chow (2002, p. 111) describes this subjectification of modern individuals as "an inherently paradoxical, frequently self-contradictory, but nonetheless effective set of give-and-take mechanisms that, once set in motion, makes it possible for the subject to function rationally in the modern world". For both Foucault and Chow, confession provides the needed technique and process to produce this subjectivity pivotal to the power relationship with the attraction.

In their book *The Confessing Society*, Fejes and Dahlstedt (2013) suggest that the practice of confession is coupled with the repressive hypothesis, since "citizens are construed as fragile and vulnerable people who need support to handle their emotions". The consequence of not conforming to this symptom of emotional deficit is punishment, deeming the individual "in denial or repressing their 'true' feelings" (Fejes & Dahlstedt, 2013, p. 2). The scholars further show that lifelong learning has been organized and structured as a power field where individuals are shaped into desirable citizens through their confessional self-reflection and self-examination in education, work, and private life. This lifelong learning enables aspects of the self, such as health and beauty, to become visible for public scrutiny, driven by the popularity and advancement of online media and social platforms.

Fandom and fans, I believe, stand on the frontline of our confessing society through their explicit experience of power as attraction. It is not a coincidence that fan scholars are suggesting fandom heralds a future for the human species in general. Jenkins (2007, p. 361), in a paper titled "The Future of Fandom", argues that the ideal human individual is "almost certainly a fan" and that fandom has become "part of the normal", suggesting "we are all fans, or perhaps none of us is". This certainty of being a fan, resonating with MacCannell's (1976, p. 1) observation that the tourist is "every modern man in general", captures how power has been mobilized to control, regulate, and manipulate individuals through its attractions. A fan is attracted to an attraction, just as the tourist materializes this process through their trip. Noting the power structure and forces that fandom and fans have harnessed, Booth (2015) proposes that fandom is the classroom of the future because, similar to lifelong learning opportunities, "one's fan identity might be the catalyst of critical intellectual shifts" after completing formal schooling. By linking fandom to schools, Booth (2015) makes a manifest claim that fans are educated, and disciplined bodies, as schools, one of Foucault's most discussed stately disciplining institutions, have been performing their duties with the production of students. For Booth (2015), fandom even excels in its discipline and regulation because there is a "momentous opportunity and responsibility to learn from and to teach fannish ardor in the classroom". In fact, we might venture to say that perhaps the ubiquitous and universal presence of fans and the fan population has made fandom one of the most sophisticated and advanced institutions/power mechanisms where power has maximized its discipline over individuals as subjective beings and regulated the fan community as a new form of population.

What makes fans and fandom particularly important to contemporary and future society is also the fan's integral, intensive, and sometimes creative use of the technique of confession. An individual fan emerges as they experience the pleasure of the attraction while making a confession detailing this experience. This turn toward the self, as "a form of production" (Chow, 2002, p. 111), suggests that the individual must be seen both to own their self and to exhibit this self repeatedly, while this process of detecting the self and exhibiting the self both take the form of confession.

In his paper exploring the relationship between confession and identity, Marshall (1998, p. 1) purported that confession is "at its base, a revelation of the self. It is marking one's boundaries of identity, particularly in how that identity is publicly displayed". In the same paper, Marshall (1998, p. 2) soon discovered that he, a cultural critic, had developed a "fan-like relationship to the exploits of Patrick Rafter, the recently recrowned US Open tennis champion". He further explores his confession and analysis of his self (Marshall, 1998, p. 2):

> Why is this a confession? Because my public self is not totally in line with my private self. Knowledge about my interest in tennis is limited: it's there but it is constrained and not easily revealed. But a fan's relationship is also somehow illegitimate for an academic, a cultural critic—at least with reference to how I present an ideal self outwards. Fandom indicates a beyond-rational affinity, a proxy relationship to activity and agency. One can make an academic claim to fandom as Henry Jenkins does so convincingly in Textual Poachers; nevertheless the presentation of the self as fan weakens the idea that the fan can achieve notoriety for his or her own work and activity. Rafter is actualised fantasy for me bundled with a resignation that I can only achieve the sensation of brilliant tennis and tennis fame through a public figure. My confession imbricates all of the elements of the public and the private. Rafter is a public figure and by association I can live vicariously through his achievements. My confession is also complexly pitched: within the field of cultural studies dealing with the biographical, the construction of subjectivity is an essential and vital practice. I can confess this detail as long as I write this preamble about the role of confession in the construction of contemporary identity.

In this case, Marshall uses his confession to construct his relationship with Rafter, as it provides him with an opportunity to reorganize the gap between his public and private self. For the scholar, fandom served as a means to fix his rational and professional academic life by building a fantasy around a public figure. Through the scholar's examination of himself – what has become public, what remains private, and how these two parts intersect with what is presented and exhibited, in this case, Rafter – Marshall navigates the power of worldly attraction. Marshall's confession of his fanship with Rafter, at the same time, remained exclusively private, as the scholar continued to reflect on his own career, profession, and lived experience through Rafter's career. In this singularity, Marshall's confession was about himself alone, as the questions he posed were specific and pressing, isolating and highlighting the pleasures and peculiarities of his own experience. This process makes the pleasure of living through Rafter and investigating Marshall himself more distinct and pronounced.

Fans have traditionally used fandom to negotiate self-identity (Dajches, 2022; Hao, 2020; McInroy & Craig, 2018; Williams, 2015). Based on Foucault's notion of power, the critical question fan studies must address is how this self/individuality has been confessed within fandom. In the following, I investigate the panda fandom as one of these power sites where the Chinese have been regulated as a confessing population, while their private and living labor has been mobilized to experiment on and sustain the economy and politics of the giant panda industry.

Specifically, I will also explore how the biopolitics of cuteness and Chinese ethnicity have introduced critical confessional mechanisms for the Chinese public to produce their selves and subjectivities.

What is an animal?

The significance of confession in Foucault's perception of power and biopolitics is that Western people have become "confessing animals". How is this confessing animal different from, or similar to, the animal in general – such as the giant panda in the panda fandom? Is a confessing animal an animal or a human? If confession has transformed the Western man into an animal, is he still different from a giant panda? If the giant panda could, say, confess, would that make it human? Is the confessional animal on par with the "natural animal"? Returning to Foucault's own statement, it seems that confession provides the critical link between the Western man and the animal, as the confession simply transforms the man into a beast. How does this transformation take place? Foucault (1990, pp. 61–62) provides the following explanation:

> The confession is a ritual of discourse in which the speaking subject is also the subject of the statement; it is also a ritual that unfolds within a power relationship, for one does not confess without the presence (or virtual presence) of a partner who is not simply the interlocutor but the authority who requires the confession, prescribes and appreciates it, and intervenes in order to judge, punish, forgive, console, and reconcile; a ritual in which the truth is corroborated by the obstacles and resistances it has had to surmount in order to be formulated; and finally, a ritual in which the expression alone, independently of its external consequences, produces intrinsic modifications in the person who articulates it: it exonerates, redeems and purifies him; it unburdens him of his wrongs, liberates him and promises him salvation . . . On the other hand, the agency of domination does not reside in the one who speaks (for it is he who is constrained), but in the one who listens and says nothing; not in the one who knows and answers, but in the one who questions and is not supposed to know. And this discourse of truth finally takes effect, not in the one who receives it, but in the one from whom it is wrested.

In this paragraph on confession, Foucault (1990) explains the power relationship between the confessor and the listener. The listener, as the dominant force, is able to extract the truth from the one who confesses. In contrast, the one who speaks is the subject of power, constrained and compelled to reveal their truths. Given this power dynamic, the confessor is reduced to an animal, compelled to make corrections, changes, redemptions, and purifications through consistent self-examination and self-exploration. The authority holding the power to listen to the confession can perform a variety of tasks, such as demanding the confession, offering prescriptions, giving appreciation, making judgments, administering punishment, and facilitating reconciliation. The complex and multiple roles that the listener can assume suggest that this authority figure is not merely an individual human being but a set of social institutions, as in the disciplinary power, and stately organizations, as in biopower, that receive, interpret, and investigate what has been confessed.

This further explanation of the use of confession shows that Foucault (1990) does indeed propose the possibility of traversing the animal–human boundary when he relates Western man to confessing animals. Instead, Foucault uses the term "confessing animal" to describe the power trap that Western man has stepped into as confession assumes the social norm and truth. This means that the confessing animal does not engage with the "natural animal" on equal ground, where unprejudiced conversations might take place, but instead highlights a similar dilemma faced by the nonhuman animal within the power structure. In this light, the confessing animal remains an anthropocentric gesture by Foucault and does not offer a platform for nonhuman beings in the process of confession.

It is this exclusion of nonhuman beings from the figuration of the confessing animal that brings us back to understanding that animals are a super-exploited means of production, or dead labor, and private property within the power structure of capitalism. The giant panda, even if she knew how to confess, does not automatically transform into a human figure, and her inability to confess confirms that the animal's participation in Foucault's power structure is not through the path of man. This suggests that the animal, as a nonhuman being, can be understood as part and parcel of the institutions through which contemporary human beings are produced. In fact, a close examination of the agency of domination in Foucault's (1990, p. 161) confession – "the one who listens and says nothing; not the one who knows and answers, but the one who questions and is not supposed to know" – suggests that, as a means of production, the nonhuman animal can exert order and control over the confessing human beings. The co-presence of the nonhuman animal with power institutions and the state, rather than with individuals, makes the nonhuman animal one of the power sites requiring and attracting confessions.

This institutionalization of the animal, this power-enthroning of nonhuman beings, I suggest, is how animals have been super-exploited in both Foucault and Marx. It does not mean that animals have been exploited more than the working class, such as the confessing man – that the animals' labor has received no compensation or wages that could at least maintain their livelihood, and the animals, in human society, often receive minimal care and welfare benefits. Rather, this super-exploitation occurs because nonhuman animals have been seamlessly integrated into the locus and site of power. In Foucault's confession, animals are burdened with the role of the authoritative figure who must listen, acquire, judge, and appreciate the confession provided by their human counterparts in their animal's silence and presence. If the Western man, through his confession, occupies the position of the animal in the power relationship, the nonhuman beings are left with a fixation through the gaze. In this light, the super-exploitation of animals in contemporary society cannot be simplified as a less-than-laborer state in which animals suffer from relentless deprivation and total exploitation. Instead, this super-exploitation is best understood as a coerced power relation involving nonhuman beings who may not have willed this power in the first place.

As one may already have questioned, how does the animal assume this power position over human beings? What tactics or strategies do animals employ to move beyond the human in general? Power, as Foucault (1990, p. 94) explains, "is not something that is acquired, seized, or shared, something that one holds on to or allows to slip away; power is exercised from innumerable points, in the interplay of nonegalitarian and mobile relations". In this proposition of power, Foucault emphasizes the importance of escaping the traditional systems of law-and-sovereign, which consider power as something one can acquire, seize, or own. Rather, power exists through its exercise. Animals in power does not mean that animals have actively sought, acquired, or seized power over human beings. Instead, animals come into power when they become part of an attraction and continue to draw attention from their human counterparts. As means of production, animals can exercise this power simply by their being – their presence in zoos invokes curiosity and interest within individuals and enforces social norms that regulate people as a population.

Animals in power also refers to the internalization of their resistance – an animal's unwillingness or indifference to the power network. According to Foucault (1990, p. 95), "this resistance is never in a position of exteriority in relation to power". Foucault suggests that resistance is not an external force opposing power from the outside but is an inherent part of the power relationship itself. This mutual interdependence between power and resistance means that resistance is produced and shaped by the very mechanisms of power it opposes. Precisely because animals have been considered the reference point for the human species, their silence, inscrutability, and irrelevance to human society become sites of resistance incorporated into their power relations. One might even venture to suggest that since animals have always been bodies resisting the network of power, this resistance allows them to occupy a place of power against the human species.

In panda fandom, the giant panda, as such, is the locus of the power–resistance relationship and a means of production super-exploited within the capitalist economic system. Suggesting that the panda is in power, in contrast to the exploited fans and their free labor, highlights how the giant panda has been integrated into the power structure and capitalist system differently from the fans. It means that the valorization of the giant panda's labor, in the fashion of respecting fans' contributions and work in the fandom (Stanfill, 2019), does not mean the animal's liberation from the power system. Rather, the different roles in the power system – fans as confessors and animals in power – mean that resistance, if there is any, should deploy different strategies through these two different sites. This point is perhaps not emphasized enough in existing studies where animal emancipation and liberation are framed through the working class's struggle for power in capitalism.

By "animal in power", I do not mean that animals have become the dominating, oppressive ruling class over human beings, but that even this power position can become a trap for animals who might not desire this role in the first place. The panda in power and its entrapment within the power structure can, I believe, be best illustrated through the emergence of panda celebrities in the panda fandom.

Panda celebrities

The giant panda is a manifestation of Foucault's (1990, p. 45) statement that "power operates as a mechanism of attraction". As an animal in power, the effectiveness and strength of the panda's power do not necessarily depend on its magnitude or centralized authority, but rather on its ability to shape and regulate behavior, knowledge, and discourse within human society. The power circulating around the giant panda allows the presence of the animal to draw out certain characteristics or behaviors from human individuals or populations over which the giant panda exerts its influence. This power through attraction distinguishes the giant panda significantly from a few domesticated animals that have integrated into human society. Unlike sled dogs, horses, or animals that perform a wide range of labor tasks for humans, the giant panda, as an attraction, rejects the top-down hierarchical position of humans. Instead, the attraction of the giant panda comes from below through commercial aestheticism, such as cuteness or the marginality of the Chinese ethnic identity. Reflecting on celebrities through Foucault, Marshall (2014, pp. 72–73) notes that the celebrity, "like sexuality, allows for the configuration, positioning, and proliferation of certain discourses about the individual and individuality in contemporary culture".

Although Marshall's discussion did not focus on animals such as the giant panda, the animal becomes a celebrity and rises into a power position because it affords a plethora of discourses with which modern individuals can structure their confessions. Animal celebrities, particularly the giant panda in this study, exercise power within a network of attraction that shapes and regulates human behaviors and experiences in a meticulous way. Following Foucault's understanding of power, we may suggest that the more diffused and ubiquitous a power appears to be, the more effectively and impactfully it will shape individuals and populations. The animal celebrity, such as the giant panda, exercises its influence over human bodies through the details and biology that the aesthetic of cuteness has readily provided.

For example, in Chapter 3, the biopolitics of cuteness mobilizes the human species through its visualization of human infantile features. Cuteness as a biopower does not position its manipulation and control of individuals through a sovereign force that monopolizes and dictates the life and death of human beings – the cuteness does not enslave or overwhelm individuals like the sublime often does. In contrast, cuteness abides with the human infant – the mundane, the detail, the necessity, and the mandatory nature of each human being – and creates a meaning and experience specifically tailored to this biological phase. In this light, the cute "comes from below" and provides "the basis for wide-ranging effects of cleavage" (Foucault, 1990, p. 94).

Coming from below, cuteness refers to a decentralized and diffused form of power. This experience of the infant and infantile cuteness is not concentrated in the hands of authority or governments. Instead, the experience of being an infant is a diffused process that is experienced throughout society, operating at multiple levels and within various institutions, such as families, kindergartens, zoos, schools, workplaces, and cultural systems. Hence, the cute draws on the microlevel dynamics of power, which emerge and focus on everyday interactions, practices, and discourses. As May (2019, p. 177) observes, if the cute speaks any language at all, "it is the language of unpindownability". Suffice it to say that the cute performs the decentralization of power to such an extent that it cannot simply be underpinned or captured by authoritative figures. The diffused presence of cuteness means that it operates through power networks that subtly shape individuals' behaviors, identities, and subjectivities in often imperceptible ways.

The panda, celebrated for its cuteness, exercises power over individuals through the dispersion and decentralization of power that emanates into various points and interactions within social structures. This exercise of power entitles the cute to an experience that May (2019, p. 187) has succinctly noted:

> Its fleeting, floating forms lightheartedly celebrate the intuition that the ordinary, no matter how familiar or insignificant, is stubbornly uncanny; and that, for all our struggles to be securely at home in the world, we are misfits living in uncertainty. In giving style to our misfittedness and lightness to our uncertainty, Cute is able to inspire a sense of freedom—above all, freedom from the tyranny of identity and power: from a human life rigidly understood in terms of clear identities and degrees of power to which it strives to give sincere and authentic expression. For, as we have seen, cute things or people are often, perhaps usually, of indeterminate power, gender, age, ethnicity, and morality. And except when the relationship between the cute object and its admirer is perverted into the grimly sadomasochistic, Cute readily mocks power—indeed, places in question the very purpose and value of power as well as the matter of who is really exercising it.

At this level of universal micromanagement and microphysical control, the cute, in contrast, is not experienced as a form of control and management. Rather, the cute inspires "a sense of freedom" and can readily "mock power". Coming from below, the cute is not a power concentrated and accumulated through a central figure whose hegemony represses and excludes the bodies under control. Instead, the cute spreads through a net-like structure that produces pleasure and experiences at the intersections among bodies, institutions, states, and organizations. This, for Foucault (1990), is the remarkable feature of power, as it exercises discipline and control through frivolity, pleasure, and the "making of life". It is not hard to detect the pleasure May has observed in the cute, as this aesthetic experience readily subverts and resists what is perceived as tyranny, identity, authority, and power. The production of life regulates pleasure within the bodily and biological processes rather than causing fear, terror, or pain within these bodies. Because of this focus on the production of life and the regulation of pleasure, Foucault's power mechanism is grounded in the principle of attraction. If we agree with Marshall's (2014) claim that the concept of affect is central to understanding the meaning and power of the celebrity in contemporary culture, we are also suggesting that celebrities are sites and focal points where Foucault's power manifests and materializes through its attraction.

The power of the cute also lies in its ability to structure resistance at different levels. On the most personal and individual level, the cute provides a sense of freedom against the tyranny of authority and power. The diffused and dispersed prevalence of the cute exercises this resistance in a perhaps most egalitarian gesture, offering individuals a similar experience. For May (2019, pp. 45–46), the cute can itself be a way to aestheticize the politics of power because it does not "only draw attention to inequalities of power; crucially it also questions their very purpose and effect". The cute provides a vantage point to examine and detect how power is exercised and performed, and this self-examination and self-investigation can readily inform possibilities to resist the dominant ways in which power operates. Ngai (2012) considers the cute an "intensification of commodity fetishism" while introducing a point of departure from the abstraction of labor through the invitation of physical interaction. Hence, the cute attempts to reclaim the tangible and sensuous appeal of the use value of commodities, which is lost in their abstract exchange value. Cuteness, therefore, embodies a dual layer of consumer fetishism, focusing on physical interaction with commodities while simultaneously obscuring their true nature as products of labor.

What makes the animal celebrity most attractive, however, is how power and resistance reach a point of balance through its reconciliation of the conflict between laborers, labor time, and the means of production. Human celebrities, as we know, perform numerous forms of labor, such as red carpet labor (Lawson & Draper, 2021), attention labor (Van Krieken, 2018), symbolic labor (Van Krieken, 2018), and emotional labor (Ferris & Ferris, 2011; Goodman & Barnes, 2011; Nunn & Biressi, 2010). The visibility and attraction of celebrities, despite the network of power, also stem from the unique values that celebrities create as workers. In contrast, I have demonstrated that even in their celebrity and stardom, animals remain the means of production in capitalist relationships. The differing roles that animal celebrities and human celebrities occupy reflect a critical gap in the power and resistance they exercise. In Marx (2000), the means of production, also termed dead labor, are consistently in conflict with laborers who sell their living labor to survive. While living labor is exploited and exchanged to generate profits, capitalists appropriate it to further accumulate dead labor as private property. Marx argues that the means of production and laborers exist in a dialectical relationship. While laborers depend on access to the means of production to work and earn a living, capitalists exploit laborers to operate the means of production and generate profits. This lack of control over both the means of production and the labor process, Marx suggests, estranges and alienates laborers from the outcomes of their labor.

With animal celebrities, however, this estrangement and alienation between the fans, as devoted free laborers, and the animals, as a unique means of production, are perhaps secondary to the pleasure and attachment the fans feel toward the animals. Fans of panda celebrities, clearly, do not enjoy the exclusive ownership of the animals experienced by capitalists, but the impact of exclusion from ownership is mitigated and compensated for as the fandom affords flexibility and more freedom in the fan's labor process. In other words, while a profit-driven capitalist venue alienates laborers through the process of labor – often characterized by the division of labor and specialization, where laborers perform repetitive and often monotonous tasks – fandom, at least on the surface, affords a different process of labor that has been termed "free labor". Terranova (2012, p. 33), in her exploration of the digital media industry, suggests that the emergence of free labor by media fans is a consequence of shifting work processes from the factory to society, where "a truly complex machine" is set in motion. Furthermore, the provision of free labor is conditioned in a complex and experimental balance between long-standing cultural and emotional aspirations for creative expression and contemporary capitalism's prioritization of knowledge as the primary contributor to value enhancement (Terranova, 2012).

Self-objectification

This devotion to free labor, I believe, also creates an intricate connection between the means of production and the laborers – a connection Marx's alienated workers could neither imagine nor perceive. This partial connection – I call it partial because it differs from "ownership" or "owning" – experienced by the fans resonates, to a certain extent, with the panda celebrity's conditions within the capitalist system. Let me revisit the anecdote of Sir Peter Scott's creation of the giant panda logo for WWF. According to Scott (WWF, 2022), the rationale behind the panda design was as follows:

> We wanted an animal that is beautiful, is endangered, and one loved by many people in the world for its appealing qualities. We also wanted an animal that had an impact in black and white to save money on printing costs.

If we substitute "animal" with "fans" or "fandom", it is surprisingly apparent that this statement still makes sense:

> We wanted fans/fandom that are (is) beautiful, are (is) unique, and one joined (loved) by many people in the world for its appealing qualities. We also wanted fans/a fandom that had an impact in capitalism to save money on labor costs.

This possibility to alternate between "fans" and "giant panda" in Scott's comment suggests that cuteness, beyond being a personal and an individual aesthetic experience, can signify a collective awareness or affect politicization to mediate tensions inherent to capitalist production. In Chapter 3, I argued that a cost-saving giant panda is, in fact, cuter than a beautiful but expensive animal. As a means of production, the cute giant panda attracts the attention of fans who devotedly consume and work for the animal. What does it mean, then, that fans and fandom can also be described as cute? Are cost-saving and free-working fans not cute? What becomes intriguing in this connection between the giant panda and the fans is that the borderline between the workers and the means of production begins to dissolve. When fans or panda fandom become cute and free workers, the term that best describes this process is objectification (Gruenfeld et al., 2008, p. 124), also termed self-objectification, defined as "seeing oneself in terms of others' goals".

Objectification involves converting human beings into thing-like entities that subjugate people as means to an end. Gruenfeld et al. (2008) suggest that, given that social connections are crucial for determining both social and economic success, the capacity to make oneself useful – particularly to influential individuals – might be an essential route for upward mobility. Moreover, the psychological costs associated with objectification may depend on the nature of the individual's own goals. Gruenfeld et al. (2008) propose that when power aligns with individuals' own purposes and attributes, objectification can even lead to self-actualization. The objectification of fans, or the fans' self-transformation into means of production, explains the affinity that panda fans share with panda celebrities within the panda fandom.

This alignment between fans and panda celebrities suggests that panda fandom is also a sphere of *biopoliticization* in which fans self-objectify. Since biopower governs human life by treating human beings as biological entities to be managed, fanning a giant panda entails objectifying and reducing the self to a biological entity subject to control. As Ngai (2012, p. 24) observes, cuteness – initially an affective response to weakness or powerlessness – is frequently "overpowered by a second feeling of manipulation or exploitation that immediately checks or challenges the first". While the manipulation and exploitation of the means of production are inherent to capitalist relationships, fandom and fanship highlight the control and organization the means of production exert over workers through the fans' self-objectification.

In this profoundly alienated experience of fandom, fans actively reject recognition through their labor, seeking instead acknowledgment through their affection and commitment. Stanfill (2019, p. 19) suggests that fans provide free labor to justify their "rejection of capitalist projects". I argue, however, that free labor is a byproduct of fandom rather than a means of rejecting capitalist logic. Instead, fans' provision of free labor results from their extreme alienation and objectification within capitalist relations – the *thingification* of fans entails that they demand recognition not for their intrinsic humanity or the individual qualities represented by personal labor but for how object-like they have become.

Indeed, we might suggest that fan labor is free in the sense that labor performed at an individual level in contemporary society holds little or no value for fans. Rather than trading their labor for meager wages as the working class historically has, fans explore objectification as an alternative to becoming essential, capitalized, and owned properties of the capitalists. The question for fans is not how they are exploited as both workers and a means of production in capitalist relations, but whether, through objectification, they can become capital and property necessary to the capitalist agenda rather than disposable laborers within the ruthless progression of capitalism. In this space opened up by reification as a means of production rather than exploitation as laborers, fans can challenge or "mock power", as cuteness plays with power relations.

Self-objectification as a means of production and the rejection of labor valorization provide avenues for fans to aestheticize panda celebrities. The giant panda Hua Hua, the superstar singled out from the species in general over the past 3 years, provides an example of the devaluation of the labor process and the valorization of the thing-like features of the panda. *Global Times* (2023) offers a brief description of Hua Hua's appeal:

> He Hua (nicknamed Hua Hua) has captivated the hearts of Chinese people thanks to her mild temperament and her rare inability to climb trees. She is also different from other giant pandas because of her white fur, rather than the more common beige color, and her chubbiness, making her highly recognizable. Hua Hua has also become popular on international social media platforms and is adored by overseas viewers.

Researchers have noted that giant pandas are excellent climbers (Brambell, 1976; Sheldon, 1937). Sasaki et al. (2005) suggest that the panda bear's hindlimbs are well-adapted for tree climbing, supported by the development of its tibia and tendon, indicating that giant pandas are genetically coded to climb trees. If Hua Hua lived in the wild, her inability to climb a tree could readily disqualify her from being considered a proper giant panda. However, it is precisely this inability to perform as a typical giant panda – that Hua Hua does not need to "labor" like one – that allows her to rise to fame and celebrity. Hua Hua captivates and attracts because she is insufficiently a giant panda, paving the way for her reification. Hua Hua's appearance – her mild temperament, chubbiness, and distinct fur – visibly sets her apart from her counterparts and validates the perception that value is inherent in the natural and concrete attributes of objects rather than in the labor process.

By highlighting the rejection of the labor process and the provision of free labor, I do not mean that labor itself has ceased to be the unique source of value. Rather, it is the extreme reification and alienation of labor – where the physical surface, the visible cuteness, and the desire to touch, grab, and hold overshadow the abstract – that allow the object and product to eclipse the labor behind them. This explains why Meng Lan, a panda celebrity boasting "a broad set of skills like climbing, digging, doing sit-ups, and feeding birds", as well as a "bold escape" from the zoo, and described as a "hardened criminal" (Global Times, 2023), achieves fame secondary to Hua Hua. While Meng Lan's impressive tricks are fascinating, this panda labor has yet to generate the same level of value, attention, or attraction in today's panda fandom as Hua Hua's presence.

Embodied gap: A panda is a celebrity but not a human

In one of the few studies devoted to animal celebrities, Giles (2013, p. 117) observes that animal celebrities "wear the badge of authenticity that is held to be so important for credible image-management". Giles furthers this by suggesting that since fans are often intrigued by the psychological puzzles posed by authenticity and the effort to uncover the "veridical self", the absence of such an authenticity dilemma makes animals less suitable as celebrities. Thus, Giles (2013) posits that animals achieve celebrity status solely due to the affective power invested in them by their fans. Similarly, Blewitt (2013) contends that animal celebrities often serve to bridge significant divides between humans and nonhumans, nature and culture, the material and the semiotic, and the organic and the technological. This mediated charisma of nonhumans tends to blur the boundaries of modern human subjects, creating opportunities to analyze nonhuman differences and the diverse agency potentials of various organisms. Furthermore, as humans advocate for, empathize with, question, communicate with, or observe these nonhuman entities, the animals often respond in ways that transcend mere reactivity, demonstrating cultural, personal, and social interactivity.

However, while the giant panda – arguably the most prominent animal celebrity in contemporary society – is affectionately celebrated for her cuteness and charisma, her status does not converge the divide between human beings and nonhuman species. Becoming a celebrity has not granted the giant panda a human-like status, nor has it prompted her treatment or valuation as a human. The fact that a giant panda can achieve celebrity status does not challenge the truth that the panda is not human. In this sense, the creation of an animal celebrity does not facilitate a more-than-human community where humans exist as part of a broader ecological and ontological system (Abram, 2012). Instead, Marxist and Foucauldian theories reveal that the giant panda's celebrification occurs within a framework of exploitation. From a Marxist perspective, the difference between human and animal celebrities lies not in fans welcoming animals solely through affection, but in the fact that animals in fandom function as a means of production, capital, or resources exploited by humans. While fans, as laborers, may develop affectionate relationships with the means of production and even objectify themselves, this affection and objectification only reinforce the distinction between the means of production and the labor force. Conversely, fans' relationships with human celebrities, who operate as either laborers or capitalists within the celebrity economy, are marked by class conflict or class awareness.

Foucauldian perspectives allow us to further question how the exploitative interactions between pandas as capital resources and fans as laborers are reconciled in the panda fandom. This inquiry positions animality as a topic parallel to Foucault's analysis of sexuality. In *The History of Sexuality*, Foucault (1990) argues that since the Victorian era, sexuality has been repressed into silence, while medical and scientific scrutiny proliferated discourses about sexuality by classifying behaviors and identities into categories of normalcy and deviance. On the surface, the animal celebrity appears to have minimized animality or the human–animal divide. Not only is animality diminished, but it also seemingly has no right to exist and is erased upon its slightest manifestation. Everyone knows the giant panda is a celebrity, which is why her animality is neither acknowledged nor treated as such. Foucault's (1990, p. 105) insight into sexuality offers a compelling analogy here:

> In actual fact, what was involved, rather, was the very production of sexuality. Sexuality must not be thought of as a kind of natural given which power tries to hold in check, or as an obscure domain which knowledge tries gradually to uncover. It is the name that can be given to a historical construct: not a furtive reality that is difficult to grasp, but a great surface network in which the stimulation of bodies, the intensification of pleasures, the incitement to discourse, the formation of special knowledges, the strengthening of controls and resistances, are linked to one another, in accordance with a few major strategies of knowledge and power.

The panda celebrity channels Foucault's notion of power, materializing how power operates and circulates between bodies. The rise of animal celebrity emerges at a historical juncture when animals have become central economic, cultural, and political entities integral to human life. This development is predicated on the production of the human–animal divide as knowledge and truth. It is precisely because the panda's animality is never questioned that the giant panda can safely assume the role of a celebrity. This celebrification of the animal justifies its exploitation, as the human–animal divide is not perceived as a bodily difference but as a pleasure experienced within the human individual. The fact that animal celebrities are not immediately conceived as natural resources for human exploitation but are instead embedded in the "making life" experience of the human species reinforces the human–animal hierarchy. It ensures that this divide is exercised by individuals and bodies rather than being dictated by laws or enforcement mechanisms.

The panda celebrity represents a historical construct where, despite the animal's rise to fame, the boundary between the giant panda and humans remains untraversed and unquestioned. Animal celebrity concerns the production of animality and the human-animal divide as historical constructs reproduced through techniques of knowledge and power. The scientific discourses surrounding the giant panda contribute to its construction as a distinctive species within human society.

Despite its celebrification, the giant panda remains a natural resource exploited by humans. The celebrification of the giant panda serves to perpetuate its capital exploitation, sustaining human–animal structural inequalities and consolidating human economic dominance over the animal.

Panda fandom operates within contemporary power devices and has been expanding at an accelerating pace. Rather than defining animals and humans, or the forbidden and the permitted, as static power positions in human–animal interactions, the panda celebrity operates through "mobile, polymorphous, and contingent techniques of power" (Foucault, 1990, p. 106). This means that the panda celebrity does not arise from the panda's animality in opposition to the fan's humanity but instead transforms the human–animal gap into the image of the panda celebrity. In this context, it is legitimate to propose that the human–animal gap has been celebrified. This celebrification of animality implies that the gap is no longer a defined link between partners or statuses but is instead extended and diffused as a form of control. The human–animal difference is not a law or rule separating the two species but pertains to "the sensations of the body, the quality of pleasures, and the nature of impressions, however tenuous or imperceptible these may be" (Foucault, 1990, p. 106).

Economically, while animals historically symbolized the circulation and transmission of wealth, the animal celebrity actualizes this economy through "numerous and subtle relays" (Foucault, 1990, p. 107) that produce and consume bodies. Animals and their symbolic values historically functioned as a "homeostasis of the social body" (Foucault, 1990, p. 107), maintaining balance, stability, and well-being within human communities through structures like domestication, symbols, and totems. The rise of the panda celebrity signifies managing and controlling populations and bodies in more intricate and pervasive ways, extending beyond mere reproduction to encompass a broad array of social, political, and cultural mechanisms. These include celebrification, education, legal frameworks, social media, and cultural narratives that collectively shape and regulate human–animal interactions and identities.

The celebrification of the human–animal divide leads to its embodiment as individual and personal experiences, making critical evaluation from a universal or collective standpoint challenging. As it becomes part of bodily experience, the issue of the nonhuman's exploited presence consistently evades individual scrutiny, rendering efforts to cross the human–animal boundary difficult. However, this complex interplay between the panda's exploitation and celebrification demonstrates how Foucault's notion of power functions as a networked force, wherein conflicts and paradoxes become intrinsic to its exercise. The giant panda embodies an exploited celebrity whose existence is integral to the ongoing economic and political agendas underpinning human survival and future prosperity. While the giant panda is elevated to celebrity status, shaping the pleasure, identity, and individuality of its fans, this affection and adoration – rooted in the fans' self-objectification – leave the human–panda difference untouched.

A panda is Chinese and loved but still . . . not a human

In a report on panda fandom, Ji (2023) notes that panda photographer Mengqi Zhou first became interested in giant pandas during a visit to the Panda Base in 1992. Zhou recalled that around 95% of visitors at the time were foreigners. Reflecting on this, Zhou stated: "Since so many foreign friends travel great distances to China to see pandas, why not use a camera to capture this and let more people see it?" This idea marked the beginning of his career in giant panda photography. What Zhou could not have anticipated was that, 30 years later, the number of visitors to the Panda Base would grow exponentially, with Chinese visitors now dominating the market. Ji's (2023) report aligns with anecdotes shared by my colleagues, who recounted that, in the 1990s, Westerners were the primary visitors to the Panda Base before its significant rise in popularity among Chinese tourists.

Zhou's comment about the start of his photography career offers insight into this shift from Western to Chinese visitors. Notably, Chinese visitors were not initially engaged panda viewers in the early 1990s. Zhou himself was not drawn to the pandas' image or allure at first. Instead, he was captivated by the sight of Westerners traveling great distances to see the giant pandas. For Zhou, the spectacle was less about the pandas themselves and more about the fascination they evoked in foreign visitors. This observation compelled him to take up panda photography, aiming to present these scenes to a wider audience – particularly to Chinese people.

An increasing number of studies explore the intersection of colonial projects and human–animal interactions. Wadiwel and Fanon (2020) argue that critically rethinking animals requires simultaneously dismantling colonial logics. However, Montford and Taylor (2020) observe that contemporary decolonizing efforts often idealize non-Western and Indigenous relationships with animals in two ways. First, non-Western peoples are frequently portrayed as living in harmony with nature and animals. For example, the concept of "the ecological Indian" (TallBear, 2000) romanticizes Indigenous peoples as inherently eco-friendly due to their supposed innate ecological wisdom. Similarly, traditional Chinese philosophy upholds harmony between humans and nature, encapsulated in the concept of "nature and man joined into one whole" (Hou, 1997). Second, non-Western cultures are often depicted as primitive or dependent on animal exploitation. Rita Laws (1994), for instance, encapsulates Native Americans in their reliance on hunting and animals for daily life. Wu (2002) contends that only the question "why do you ask?" can unpack the ambiguities and biases in Western perceptions, such as the stereotype of Asians consuming dogs.

Panda fandom, as Zhou's memory illustrates, resists both the romanticization of harmonious fan–panda relationships and the reductive views of the Chinese as mere consumers of pandas. Instead, it exists at the intersection of Western gaze and its impact, representing a complex negotiation between world cultures. In this case, it is strongly influenced by the emergence of modern Chinese individuals. Zhou's engagement with panda photography began with the profound impact of witnessing Westerners' fascination with the giant pandas – seeing foreigners captivated by the pandas inspired him to share this image with fellow Chinese. The visual power inherent in the giant panda attraction, which marks the genesis of contemporary "panda fandom", suggests that spectatorship and visuality are central to understanding this evolving exchange.

Chow's (1995) study of Lu Xun, a pivotal figure in modern Chinese literature, offers a compelling example of how visuality has shaped Chinese individuals' experiences of modernity. Lu Xun recounted a transformative moment in 1904 while studying medicine in Japan: Watching a movie scene where a Chinese fellow's beheading was passively observed by an indifferent group of Chinese onlookers. This scene profoundly affected Lu Xun, prompting him to abandon his medical studies at Sendai Medical School and dedicate himself to uplifting the Chinese spirit through literature. Although the imagery in their experiences differs, the pivotal moments in Lu Xun's transformation into a national writer and Zhou's decision to pursue panda photography share a common thread in the power of visuality. Chow (1995, p. 6) moves beyond traditional interpretations of Lu Xun's reaction, suggesting it was not only the apathy and powerlessness of the onlookers that moved him but also. . .

> the process of magnification and amplification that is made possible by the film medium, which, as it were, makes the spectacle spectacular, the demonstration monstrous, and thus underscores the significance of a technologized visuality. Lu Xun's reaction, we might say, is not simply focused on the victim about to be executed or simply on the passive observers; more important, it is an index to the relationship between visuality and power, a relationship that is critical in the postcolonial non West and that is made unavoidable by the new medium of film.

Zhou's photography of the giant panda offers compelling testimony to the power of visuality. On the one hand, Zhou's interest in photographing giant pandas and Western visitors centered on the foreigners' attraction to the pandas' cuteness. It was an effort to capture and convey the devotion and fascination that Western visitors demonstrated toward the giant panda. On the other hand, Zhou's career in panda photography – much like Lu Xun's writing – was an exploration of the relationship between visuality and power. Positioned as both a spectator and an observer, Zhou documented the spectacle of foreign visitors drawn to the giant panda while becoming increasingly aware of how the panda, as a symbol of his "Chineseness", was itself closely scrutinized by Western audiences. Witnessing this intimate relationship between visuality and power inspired Zhou to "educate" and "change" his fellow Chinese. From Zhou's perspective as a photographer, what he "sees" and "discovers" is not merely the cuteness of the giant panda or the apparent interest of Westerners, but the direct, raw power of the visual medium itself. The panda–Western spectacle, as captured in Zhou's work, intensifies the pleasure inherent in the cuteness materialized through the giant panda. This pleasure juxtaposes the affection tied to the panda's cuteness with the gratification of being recognized and identified through the animal's Chineseness.

A significant distinction between Lu Xun and Zhou lies in their respective reactions to the scenes they encountered. For Lu Xun, the moment of seeing an apathetic Chinese audience passively watching the execution of a Chinese fellow on film was a bitter awakening to the weakness and backwardness of his people and nation. This experience motivated him to revolutionize and uplift his country. Witnessing such indifference and powerlessness resonated with China's "Century of Humiliation" (1839–1949), a period marked by national decline and subjugation (Callahan, 2004; Cohen, 2002; Kaufman, 2010; Wang, 2014). In contrast, Zhou's observation of the Western–panda spectacle fostered an emerging sense of national pride associated with the giant panda's image (Lindburg & Baragona, 2004; Ma et al., 2016; Songster, 2018). Yet, Zhou's interest in Western visitors' fascination with the giant panda must also be understood within the context of China's educational campaigns, which have often emphasized the narrative of national humiliation at the hands of Western powers (Callahan, 2009). Thus, Zhou's panda photography can be seen as an extension of Lu Xun's determination to revolutionize and change the nation. As Callahan (2009) observes, modern Chinese subjectivity reflects a "pessoptimistic" duality – simultaneously characterized by growing confidence in national strength and lingering insecurity and vulnerability. For Zhou, capturing Westerners' attraction to the giant panda represented a significant effort to contribute to this evolving subjectivity uniquely identified as "Chinese".

A further question worth considering is the specific message Zhou sought to convey to his fellow Chinese through his panda photography career. Was it the cuteness of the giant panda that Zhou aimed to highlight? Was it the spectacle of such a captivating animal? Was it the Western audiences' fascination and admiration for a Chinese animal? Or was it the powerful experience of being seen, particularly by Western onlookers? Perhaps it was an attempt to enable China and its people to engage with the global stage through the power of visibility. These questions, while complex, are critical to understanding Zhou's lifelong devotion to his craft. It could also be argued that Zhou's photography was an attempt to provide answers to these questions. However, the shared thread between Zhou and Lu Xun's career choices lies not only in their transformation through the power of visuality upon encountering "China" as a spectacle but also in how visuality functions as a trap for those who experience both seeing and being seen.

According to Chow (1995, p. 9), Lu Xun's self-awareness, achieved through the individuality of the Chinese, is not "only a matter of watching 'China' being represented on the screen; it is, more precisely, watching oneself—as a film, as a spectacle, as something always already watched". The power of visuality carries a "certain violence" (Chow, 1995, p. 9) that resides in the dynamic between the act of watching and the act of being watched, creating a split between the observer and a memorable image. Chow (1995) argues that Lu Xun, rather than adopting the supposed radical response to the power of visuality, reconciled this violence by returning to traditional practices, specifically Chinese literacy. Chow's study suggests that for those subjected to spectacle– such as the Chinese – the solution is not to deny, radicalize, or reject visuality's powerful influence. Instead, what is seen becomes the materiality of the individual. Lu Xun's writing career, therefore, emerged as an act of identification with his Chineseness, which simultaneously served as the source, inspiration, and technique of his work – an "obsession with China" (Chow, 1995, p. 23). Similarly, Zhou's engagement with giant panda photography can be understood as a medium through which he aligns himself with the panda and the Chineseness it represents. This obsession with the panda – an extension of China – also defines panda fandom and its growing presence across the nation.

Fanning the giant panda, as reflected through Zhou's lens and the perspective of the animal's fans, creates a visual space in which Chineseness both observes and is observed. Embedded in this visuality is an intertwining obsession with national identity and tradition alongside an acute awareness of the gaze and presence of Western audiences. For Chinese fans, observing the giant panda becomes a Westernizing experience that is nonetheless enacted and embodied through Chinese bodies. The power structure shaping this Chinese body embodies the violence inherent in the Chinese–Western gap. For Zhou, the significant insight gleaned from observing Western visitors' fascination with the giant panda was the realization that the giant panda – and the Chineseness it symbolizes – could be the subject of intense, pleasurable observation. Zhou's investment in the panda's image parallels Lu Xun's turn to Chinese cultural tradition for inspiration. As Chow (1995) suggests, this process of "becoming Chinese" through visuality necessitates the construction of the Chinese as the other and the exhibition of this otherness on a global stage.

To describe panda fandom as a Westernizing experience implies that while the Chinese – panda fans or otherwise – may share an ethnic and cultural identity with the animal, their relationship with the giant panda cannot escape the capitalist exploitation and anthropocentric domination perpetuated by Western cultures on nonhuman animals. While the giant panda may be recognized as Chinese, the animal is confined to her animality within the framework of her Chineseness. It can be said that the nation's identification with the animal – an intensified and reinforced image of the panda's Chineseness – lays the foundation for the animal's transformation into a symbol of Chinese tradition and a cultural relic, enabling further exploitation and deprivation.

In this chapter, I have examined panda fandom through Foucault's concept of biopower, Marx's analysis of capitalism, and the role of visuality in constructing national identity. These perspectives converge on the conclusion that the giant panda, within the realm of panda fandom, has been and will likely remain a super-exploited means of production, sustained by power techniques and apparatuses such as confession, objectification, and ethnicization. This conclusion underscores the urgency of further exploring and analyzing panda fandom. In the next chapter, I turn to interviews conducted with panda fans at the Panda Base to investigate whether fandom and fan discourses can serve as platforms for subverting and restructuring the animal's exploitation and to explore specific avenues for fostering subversion and resistance.

6

PANDA FAN SKETCH

Method and field reflection

This chapter establishes the link between the theoretical foundation discussed in the previous chapters and the empirical fieldwork, where I engaged and interacted with panda fans. Transitioning from the theoretical backbone to empirical observations represents a significant shift, which this chapter is designed to facilitate. However, this transition does not imply that the earlier chapters were devoid of evidence from the field – historical myths, children's stories, documentation, personal narratives, and news reports were abundant in the first part of the book. Rather, it highlights that the data or empirical evidence used previously was drawn from historical records still accessible for this study.

In contrast, the data presented in the following chapters were purposefully collected and systematically organized for specific research objectives. Because of this additional layer of intentional design, the datasets in these chapters appear richer and more diverse than the sources used earlier. This chapter seeks to provide a seamless transition between the two approaches, enabling these seemingly distinct foundations to inform, connect, and reinforce each other. Here, I present an overview of the panda fandom, reflecting on the methods of data collection, insights from the field, and some initial findings from the research.

Method and reflection

Interviewing panda fans was one of the delights of my working days from July to December 2023. When research or administrative tasks were not pressing, I would take a 10-minute walk after lunch from the staff-only workspace halfway up the hill to the Young Giant Panda Villa at the foot. Crossing a wooden boardwalk, a small fish pond, and a bamboo forest, I slowed my pace as I approached Hua Hua's enclosure, blending with the tourist crowds. At the enclosure, I encountered

DOI: 10.4324/9781003511229-7

the long queue of tourists waiting for Hua Hua, who typically took her midday siesta between 12:00 and 13:30. Like other areas at the Panda Base, tourists were directed into zigzagged queues by temporary fences. Standing at the tail of the queue, I noticed a sign informing visitors that the waiting time was approximately 1–1.5 hours. This often prompted some tourists to leave for other panda enclosures upon seeing the queue's length and the notice board. Such long queues were an everyday occurrence at the Panda Base.

My task, starting at the tail of the queue, was to approach individuals who might identify as panda fans. The strategy was simple: After explaining my purpose and obtaining consent, I would queue along with the fan while conducting the interview. This approach proved successful after some trial and error across different sites at the Panda Base. Visitors willing to dedicate their time to see Hua Hua were often fans, and queuing alongside them reduced the chance of rejection due to time constraints. I also found that wearing the Panda Base uniform validated my identity and helped build rapport with interviewees. Over time, frequent visitors to Hua Hua's enclosure became acquaintances; we greeted each other and occasionally exchanged small talk about the pandas. These fans also assisted in snowball sampling, introducing me to other members of the fan community.

I conducted semi-structured interviews using Wenjuanxing, a questionnaire platform. The first section of the questionnaire collected demographic information, while the second featured ten open-ended questions that guided my conversations with the fans. Interview lengths varied depending on factors such as rapport, the fan's personality, and their interest in the topics. Most interviews lasted between 30 minutes and 1.5 hours. All conversations were recorded after participants signed informed consent forms prior to completing the survey and interview.

Chasing fans who hurried to get ahead in Hua Hua's queue became an amusing part of our routine. My research assistant and I often joked that we were fans of fans. While panda fans ran after their beloved animals, we found ourselves running after the fans. And I don't mean "running" literally – but we kept pace with their urgency. During a seminar, I shared a video of panda fans dashing toward Hua Hua in a daily morning competition to secure a better spot. My European colleagues were bemused and remarked they would love to join the fans, exclaiming, "I could never do that in Europe!"

As an employee of the Panda Base, I was encouraged to maintain a professional approach in interactions with fans, emphasizing the importance of utilizing official communication channels for fan engagement. This guidance recognized the passionate nature of the fan community, which actively engaged with and expressed interest in the welfare of the pandas. Their enthusiasm occasionally translated into inquiries and feedback regarding care practices, which the Panda Base, as a leading research institution dedicated to giant panda conservation, addressed with scientific rigor and professionalism. The established communication protocols ensured that fan concerns were handled effectively and systematically, reflecting the Panda Base's commitment to maintaining high standards in research and animal welfare.

I generally followed this guidance. On one occasion, as I passed through a staff-only area, a young woman approached me, asking if I was part of the management team. She appeared deeply concerned about the pandas and seemed eager to speak with someone who could address her questions. Intrigued by her dedication, I asked a few questions of my own to learn more about her motivations. She quickly discerned that I was not in a managerial role and, with clear determination, shifted her focus to finding the appropriate person to address her concerns. Although our conversation was brief, I admired her persistence and passion. Reflecting on the encounter, I regretted missing an opportunity to further understand her perspective, as she exemplified the fervent dedication of many fans I had observed during my time at the Panda Base.

As a social scientist, my investigation of panda fans was grounded in the belief that the research findings would contribute to the scientific community and support the scientific activities at the Panda Base. My work aimed to show that areas such as tourism studies, the concept of cuteness, local communities, and people's preferences for animals are valid research topics within the Panda Base, which has traditionally focused on giant panda biology, reintroduction programs, and breeding techniques. Once established, my research gave me access to the panda fan community. Instead of keeping a distance, I approached the fans as part of my fieldwork and data collection process.

However, my affiliation with the Panda Base also meant that my contact with the fans was somewhat complex. On the one hand, I needed to be mindful of the organization's image and management rules, especially in interactions with fans who might be upset or critical. At the same time, my affiliation with the Panda Base helped me build trust with interviewees. Identifying myself as a social scientist at the Panda Base was crucial for gaining approval to conduct interviews. On the other hand, my academic training had made me aware of the "power relationships" between researchers and the communities they study (Walsh, 1998). In the field, researchers often start as "strangers" or "outsiders" to the group being studied (Alan Fine & Hallett, 2014). The use of what Robertson (2002, p. 787) calls the "trope of childhood" can sometimes help level the power imbalance between researchers and the people they work with. My position was not that of a naive outsider entering an unfamiliar field. The established power dynamics between the Panda Base and the fan community influenced my approach. While the Panda Base affiliation was helpful, it also posed challenges. For example, some fans doubted my motivations, asking questions like, "I don't believe the Panda Base would want to know about us" or "Why are you doing this for the Panda Base?" I addressed these doubts by reaffirming my role as a social scientist: "Yes, I know. Maybe the Panda Base does not, but I do, as a scientist".

This combination of Panda Base affiliation and my role as a social scientist provided flexibility and opportunity for conducting my fieldwork. I was motivated to explore questions that would help deepen our understanding of the panda fandom, such as: What drives the passion of panda fans? What motivates their devotion? Do they have a solid understanding of pandas, or are they simply captivated by their charm? What influences their strong attachment to the Panda Base and the pandas?

What compels them to dedicate so much of their time and energy to the giant panda? Can the behavior of panda fans be viewed as "normal" within the context of fandom? Entering a field that had not been extensively studied before offered significant freedom, especially in a setting like the Panda Base, where social scientific research is a relatively new development. Striking the right balance between scientific exploration and respect for the existing context was essential in addressing these questions.

The length of my fieldwork allowed me to establish relationships with fans, so meeting them in the queue became more than just a casual greeting – it also carried a sense of responsibility: What could I, as a researcher, do for the panda fandom? What was my role within this fandom? My previous research in areas such as feminism, ethnicity, power dynamics, animal studies, and aesthetics, whether in tourism contexts or not, pointed me toward participatory ethics as a potential framework for addressing these questions. According to Kotzé (2002, p. 21), participatory ethics

> accept(s) that knowledge and research as a means of creating, constructing or discovering new knowledges are not innocent, but do have real life effects. Participation of everyone with any interest in the research process and results, and their possible effects, consequences or applications, becomes a key ethical commitment for all research.

My task when conducting interviews with panda fans was not, from the perspective of participatory ethics, merely a knowledge-creation process driven by my role as both a researcher and a Panda Base employee. Rather, it was a process that required my active participation and understanding. I began to realize that the question "What can I do for the panda fans and the panda fandom?" led to deeper, more meaningful conversations with interviewees than the question "What is the panda fandom, and who is a panda fan?" My role, I soon understood, was not only to collect data from fans but also to engage in the co-creation of the panda fandom, to which I, too, had to contribute. Identifying what contributions or knowledge I could bring to the panda fandom became the central motivation behind initiating this book project. Of the 12 open-ended questions I discussed with panda fans, the question, "How do you envision your future with the giant pandas?" yielded the most consistent responses. In general, panda fans did not seem to have long-term visions for the panda fandom, despite their current passion and enthusiasm for the animal. I believe this is one of the key tasks and contributions a social scientist can offer to her research field and community.

Overview of the participants

Table 6.1 provides a summary of the demographic information of the interviewees. The median age of the participants was 30, with a range from 17 to 69. To ensure diversity and inclusion in the dataset, senior citizens (aged 55 and above) were specifically approached during the fieldwork. A higher number of women (38 participants, 62%) participated in the interviews. More than half of the interviewees were either in full-time employment or working as freelancers. Eleven participants had an education level below a college degree. All interviewees, in their self-assessment, identified themselves as panda fans. Following giant panda-themed social media accounts appeared to be the norm, with only 15 fans reporting they were not particularly concerned with these platforms. Pseudonyms have been assigned to help identify participants, as shown in Table 6.1.

Researchers in fandom studies have proposed a spectrum theory to measure fans' level of engagement in their activities (Agrawal et al., 2018; Santos et al., 2019; Stander & De Beer, 2016). Based on the fieldwork, I categorize fans who visited the Panda Base more than 12 times in the past 2 years as "dedicated fans" and others as "junior fans". A total of 20 dedicated fans were identified. While I recognize that social media engagement (Rahman et al., 2016; Schultz, 2017) and self-identification (Lozano Delmar et al., 2018; Trail et al., 2017) are often used to locate fans on the engagement spectrum, this simplified dichotomy stems from my personal interactions and observations of the participants. As mentioned, several fans had become acquaintances at the Panda Base, and their frequent presence at the panda enclosures was the most direct indicator of their commitment to the fandom. Dedicated fans were more likely to live in Chengdu, with only three of them residing in other provincial capitals. Interestingly, not all dedicated fans considered their devotion absolute, with some expressing a more partial attachment (e.g., "maybe yes").

Table 6.2 presents a summary of other survey questions included in the study to provide further insight into the participants. Results in Table 6.2 should not be generalized for fans in the panda fandom because participants were purposefully approached and selected in this dataset rather than randomly sampled across the visitor population. Notably, while the panda fandom can be seen as a loose community with only 18% participants stating their relationship or membership with the community, the impact of panda celebrities/stars (77.9%) was prominent in the fandom, meaning fans were attracted to individual panda celebrities rather than the giant panda as a species. The influence of social media (68.9%) and encountering the giant panda (63.9%) were two main reasons drawing people into the panda fandom, which resonated with meeting the giant panda as the primary visiting purpose (82.0%) of the panda fans. More than half of the participants declared themselves to be absolute panda fans (57.4%). Mostly likely, fans considered pandas their friends (41.0%) and family (16.4%), while pandas often serve to boost fans' mood (44.3%) and improve their conservation awareness (36.1%). Photos and videos (72.1%) were most frequently used to document fans' interactions with the giant pandas, while about 18.0% participants reported a daily documentation. Sharing the panda documents (75.4%) was popular for panda fans.

Based on the survey results, panda fans come from a diverse range of age groups, educational backgrounds, and occupational fields, with notable representation from Chengdu and other major cities. The level of fan dedication varies, and there is a significant presence on social media platforms. The panda fandom is characterized by its loose structure but is strongly influenced by social media and panda celebrities. Fans demonstrate deep emotional connections to

TABLE 6.1 Demographic Distribution of Beginner–Senior Panda Fans.

No.	*Pseudonym*	*Sex*	*Age*	*Occupation*	*Education*	*Place of Residency*[*]	*Fan Self-evaluation*	*Social Media Following*	*Dedicated/ Junior Fans*
1	Li Mei (李美)	F	17	Student	High school	Chengdu	Yes	No	J
2	Li Jun (李军)	M	18	Student	Bachelor	Chengdu	Absolutely	Yes	J
3	Zhang Wei (张薇)	F	27	Unemployed	Postgraduate	1st tier	Absolutely	Yes	J
4	Wang Fang (王芳)	F	26	Student	Postgraduate	2nd tier	Maybe	Yes	J
5	Chen Jia (陈佳)	F	28	Full-time	Some college	3rd tier	Yes	No	J
6	Liu Yan (刘艳)	F	19	Freelance	High school	3rd tier	Absolutely	Yes	J
7	Yang Min (杨敏)	F	28	Full-time	Some college	1st tier	Absolutely	Yes	J
8	Zhang Lei (张磊)	M	52	Freelance	Some college	2nd tier	Maybe	No	J
9	Zhao Ling (赵玲)	F	25	Full-time	Some college	3rd tier	Absolutely	Yes	J
10	Huang Hua (黄华)	F	34	Freelance	Bachelor	1st tier	Absolutely	Yes	J
11	Wang Qiang (王强)	M	25	Freelance	Some college	Chengdu	Absolutely	Yes	D
12	Zhou Lan (周兰)	F	28	Full-time	Some college	1st tier	Absolutely	Yes	J
13	Chen Jie (陈杰)	M	65	Retired	High school	Chengdu	Yes	No	J
14	Liu Wei (刘伟)	M	65	Other	High school	3rd tier	Absolutely	No	J
15	Wu Ying (吴颖)	F	65	Retired	Bachelor	1st tier	Absolutely	Yes	J
16	Xu Fei (许菲)	F	21	Student	Bachelor	3rd tier	Absolutely	Yes	J
17	Yang Tao (杨涛)	M	29	Full-time	Bachelor	Chengdu	Absolutely	Yes	D
18	Sun Li (孙丽)	F	40	Freelance	High school	2nd tier	Absolutely	Yes	J
19	Zhao Ming (赵明)	M	28	Other	High school	Chengdu	Yes	Yes	D
20	Ma Juan (马娟)	F	40	Full-time	Bachelor	Chengdu	Absolutely	No	D
21	Zhu Xi (朱希)	F	55	Retired	Bachelor	Chengdu	Absolutely	Yes	D
22	Guo Na (郭娜)	F	22	Student	Bachelor	3rd tier	Yes	Yes	J
23	He Juan (何娟)	F	22	Student	Bachelor	3rd tier	Maybe	Yes	J
24	Gao Yu (高瑜)	F	69	Retired	High school	3rd tier	Absolutely	No	J

(*Continued*)

TABLE 6.1 (Continued)

No.	*Pseudonym*	*Sex*	*Age*	*Occupation*	*Education*	*Place of Residency**	*Fan Self-evaluation*	*Social Media Following*	*Dedicated/ Junior Fans*
25	Lin Qi (林琪)	F	68	Retired	High school	2nd tier	Absolutely	No	J
26	Huang Wei (黄伟)	M	68	Retired	Bachelor	2nd tier	Absolutely	Yes	J
27	Shen Lan (沈兰)	F	60	Retired	Some college	2nd tier	Absolutely	Yes	J
28	Zhou Yong (周勇)	M	52	Full-time	Bachelor	2nd tier	Yes	No	J
29	Liang Mei (梁美)	F	30	Full-time	Bachelor	2nd tier	Absolutely	Yes	J
30	Wu Jun (吴军)	M	23	Unemployed	Bachelor	2nd tier	Yes	No	J
31	Xu Bin (许斌)	M	31	Full-time	Bachelor	Chengdu	Absolutely	Yes	D
32	Sun Hao (孙浩)	M	30	Unemployed	Bachelor	1st tier	Absolutely	No	J
33	Ma Chao (马超)	M	30	Other	Some college	Chengdu	Yes	Yes	D
34	Song Ya (宋雅)	F	27	Freelance	Bachelor	3rd tier	Absolutely	Yes	J
35	Zhu Peng (朱鹏)	M	42	Other	High school	Chengdu	Absolutely	No	D
36	Deng Ping (邓萍)	F	40	Full-time	Postgraduate	Chengdu	Absolutely	Yes	J
37	Peng Hui (彭慧)	F	29	Unemployed	Some college	Chengdu	Absolutely	Yes	D
38	Tang Xue (唐雪)	F	31	Full-time	Bachelor	Chengdu	Absolutely	Yes	D
39	Han Xia (韩霞)	F	43	Freelance	Bachelor	3rd tier	Absolutely	Yes	D
40	Guo Ming (郭明)	M	25	Full-time	Bachelor	1st tier	Maybe	No	J
41	He Dong (何东)	M	27	Full-time	Bachelor	Chengdu	Maybe	Yes	J
42	Gao Lei (高磊)	M	28	Other	High school	2nd tier	Yes	Yes	D
43	Lin Feng (林峰)	M	29	Full-time	Some college	Chengdu	Maybe	Yes	D
44	Feng Jing (冯静)	F	30	Freelance	Bachelor	Chengdu	Absolutely	Yes	D
45	Dong Mei (董美)	F	36	Retired	Some college	Chengdu	Absolutely	Yes	D
46	Shen Long (沈龙)	M	29	Freelance	Bachelor	Chengdu	Absolutely	Yes	D

(Continued)

TABLE 6.1 (Continued)

No.	*Pseudonym*	*Sex*	*Age*	*Occupation*	*Education*	*Place of Residency**	*Fan Self-evaluation*	*Social Media Following*	*Dedicated/ Junior Fans*
47	Ren Hui (任慧)	F	28	Freelance	Bachelor	Chengdu	Yes	Yes	D
48	Lu Qian (卢倩)	F	36	Freelance	Postgraduate	Chengdu	Absolutely	Yes	D
49	Liang Chen (梁晨)	M	25	Student	Bachelor	2nd tier	Yes	Yes	J
50	Jiang Yun (姜云)	F	53	Retired	High school	2nd tier	Maybe	Yes	D
51	Cai Hua (蔡花)	F	20	Student	Bachelor	3rd tier	Yes	Yes	J
52	Du Juan (杜娟)	F	40	Full-time	Bachelor	1st tier	Maybe	Yes	J
53	Ye Qiu (叶秋)	F	28	Freelance	Bachelor	1st tier	Maybe	No	J
54	Cheng Li (程莉)	F	37	Full-time	Bachelor	2nd tier	Maybe	Yes	D
55	Pan Lin (潘琳)	F	30	Full-time	Bachelor	1st tier	Absolutely	No	J
56	Xiao Mei (肖梅)	F	28	Full-time	Postgraduate	2nd tier	Yes	Yes	J
57	Yuan Yue (袁悦)	F	34	Full-time	Postgraduate	1st tier	Maybe	Yes	J
58	Hou Yi (侯怡)	F	42	Freelance	Postgraduate	1st tier	Absolutely	Yes	J
59	Song Yang (宋阳)	M	42	Freelance	Bachelor	1st tier	Absolutely	Yes	J
60	Meng Xue (孟雪)	F	25	Student	Postgraduate	2nd tier	Maybe	Yes	J
61	Deng Lin (邓林)	M	28	Full-time	Postgraduate	1st tier	Yes	Yes	J

**Note:* First-tier cities refer to Beijing, Shanghai, Guangzhou, and Shenzhen. Second-tier cities refer to provincial capitals other than Chengdu. Third-tier regions refer to the rest.

TABLE 6.2 Summary of Survey Questions.

Measure	*N*	*%*
Visiting frequency in the last 2 years		
1–2	35	57.4
3–12	6	9.8
>12	20	32.8
Visiting purposes (multiple choice)		
Meeting the panda	50	82.0
Holiday	19	31.1
Hanging out with friends	8	13.3
Conservation education	2	3.3
Other	5	8.2
Do you consider yourself a panda fan?		
Maybe yes	12	19.7
Yes	14	23.0
Absolutely yes	35	57.4
How frequently do you read panda-themed social media posts?		
I am not following any	15	24.6
Occasionally	11	19.0
Weekly	12	19.7
Daily	23	37.7
Are you interested in one or a few panda celebrities?		
Yes	47	77.9
No	14	23.0
Are you a member of panda fan community?		
Yes	11	18.0
No	50	82.0
How long have you been a panda fan?		
Less than a year	23	37.7
1–2 years	19	31.1
3–5 years	6	9.8
More than 5 years	13	21.3
Why did you become a panda fan?		
Influence from my family or friends	12	19.7
Social media	42	68.9
Conservation education	9	14.8
I met the giant panda	39	63.9
Other	7	11.5
How do you perceive your relationship with the panda?		
Family	10	16.4

(*Continued*)

TABLE 6.2 (Continued)

Measure	*N*	*%*
Friend	25	41.0
Idol	7	11.5
Visitor attraction	9	14.8
Other	10	16.4
What does the giant panda mean to you? (Multiple choice)		
My spiritual animal	10	16.4
Mood booster	27	44.3
Improve my conservation awareness	22	36.1
Other	2	3.3
Do you document your panda encounters? (Multiple choice)		
No	15	24.6
Photos or videos	44	72.1
Social media	32	52.5
Other	5	8.2
How often do you document?		
I don't write	15	24.6
Occasionally	21	34.4
Weekly	14	23.0
Daily	11	18.0
Would you share what you have created?		
Yes	46	75.4
No	15	24.6
My level of intimacy with the giant panda (6-Likert scale; 1 = not at all related, 2 = not related, 3 = slightly not related, 4 = slightly intimate, 5 = intimate, 6 = very intimate)		
Mean = 4.82; median = 5.00; SD = 1.113		

pandas, often viewing them as friends or family, and are highly engaged in documenting and sharing their interactions with the animals. Despite the loose organizational structure, the dedication of many fans is evident, with a substantial number identifying strongly with the community and contributing to conservation awareness.

In total, 61 interviews with panda fans were recorded, resulting in a 392,000-word transcription. A research assistant assisted in converting the recordings into transcriptions between November 2023 and February 2024. All interviews were conducted in Mandarin, and the transcriptions were also in Mandarin. The 12 open-ended questions that structured the interviews are as follows:

1. How did your story with the giant panda begin?
2. Who are your panda idols?

3. What does it mean to be a panda fan?
4. How do you perceive time spent with pandas?
5. How do your families or friends support your fan activities?
6. Have you joined fan communities?
7. Do you think panda fans have received social support or recognition?
8. How are celebrity pandas, captive pandas, and wild pandas different from or similar to each other?
9. What do pandas think of panda fans?
10. Ho do you envision your future with the giant pandas?
11. Do you share similarities with the giant pandas?
12. How can you contribute to giant panda conservation as fans?

The questions in the semi-structured interviews provided a framework for the discussions while also allowing flexibility to explore additional topics (Madon, 2022). Probing questions, based on the fans' responses, enabled an open dialog and allowed for deeper exploration of their ideas. As the interviewer, my understanding of the field developed throughout the process, helping uncover relevant or unexpected themes and details that emerged during the conversations. Using NVivo 14, my research assistant and I coded the transcriptions, which resulted in a theme tree that covered the topics discussed across the interviews.

Drawing from the theoretical framework, three central themes – biopower, capitalism, and nation-state – emerged as the main branches supporting the structure of the panda fandom. In the theme of biopower, I examine the "making of life" for panda fans through their interactions with the giant panda in their everyday lives. Specifically, I focus on how participants construct cuteness and its aesthetic experience as a form of biopower, which creates boundaries between the human species and the giant panda species, as well as shaping knowledge and truth about the giant panda and regulating the emotions, bodies, and feelings of panda fans. The theme of nation-state arises as interviewees discussed the giant panda's national and symbolic significance. I explore how panda fans contribute to the creation of the notion of "Chineseness", the formation of a Chinese identity within the context of China–West relations, and how the idea of nationhood is embodied in the image of the giant panda. Under the theme of capitalism, I analyze how fans perceive a balance between their work and personal lives through their interactions with the giant panda, the psychological benefits and well-being derived from these interspecies connections, and fans' exploration, exploitation, and resistance to the giant panda's super-exploitation within the fandom.

The interplay among capitalism, the nation-state, and biopower is evident in the ways modern Chinese society organizes, regulates, and manages the panda fandom population. The panda fandom is governed as a healthy and productive community that supports the capitalist economic exploitation of the giant panda, while Chinese and fan identities are shaped to reinforce the power structures and mechanisms within the panda fandom.

7

LOVE PANDA, LOVE MY NATION

"Only China has pandas"

> "It means . . . I feel that our country has giant pandas. And regarding the protection of pandas, it's really perfect. They are really dedicated to protecting this national treasure of ours. And then, it makes me feel proud, because our country has pandas, this aspect."
>
> (Zhou Lan, F, 28, junior fan)

When asked about how her connection with the giant panda began, Zhou Lan's first statement highlighted the panda's uniqueness to China. Zhou recalled meeting a giant panda at a local zoo during her childhood and instantly falling in love with the animal. A resident of Guangzhou, she pointed out that places like Chimelong Paradise, Hong Kong Ocean Park, and Shenzhen Zoo were among the few locations where she could encounter a giant panda in the following years. For Zhou, her love and devotion to the panda, despite meeting her only once, was not a fleeting romantic notion but an authentic and deeply rooted expression. She was proud that the giant panda was a species uniquely tied to China, a treasure found only in her home country. Zhou's enduring, deep, and immediate affection for the giant panda was not an isolated sentiment but a shared experience among many panda fans. He Dong (M, 27, junior fan) expressed a similar sentiment, stating:

> So, I think that for many Chinese people, the giant panda is not just a recent sensation that has appeared before our eyes, but rather a kind of connection from childhood to adulthood, whether it's a bond or simply a curiosity about it. Moreover, the giant panda is truly adorable and loved by everyone. Basically, during childhood, once someone sees it, they will love it for a lifetime. I myself saw a giant panda at the Shanghai Zoo in my childhood, so I have always liked it.

DOI: 10.4324/9781003511229-8

He Dong understood that the love for the panda is, in a way, an inevitable experience for the Chinese. Once a Chinese person sees a giant panda in the zoo, that connection becomes a bond lasting a lifetime. The same was true for Lin Qi (F, 68, junior fan), who admitted that she didn't initially feel a strong attachment to the panda as a national treasure. However, over time, her feelings evolved, and she came to embrace the panda as an integral part of her identity and connection to China.

In the past, when people called it a national treasure, I wasn't really . . . well, I didn't like it that much. Later on, I started to see it as an ambassador, and I grew to like it more and more. Because you know . . . well . . . because of the giant panda, I started to like it as an ambassador. I would tell my grandson, 'The giant panda is an ambassador when it goes abroad and so on and so forth.' I would tell him, It's a panda, but it's also a part of humanity. It's essential to nature and also carries a very important responsibility for our country.

Lin Qi's love for the giant panda, though not initially a love-at-first-sight experience like that of Zhou Lan and He Dong, gradually developed into a deep and intense affection. She came to view the panda as an ambassador bridging China with the world, and passed this understanding on to her grandchild, highlighting the giant panda's vital role in fostering transnational communication. For Lin Qi, the panda was not only an animal connecting different cultures but also a symbol linking people to nature.

The nationalist passion that the giant panda represents is even more explicit and foundational for Huang Wei (M, 68, junior fan; YG refers to the researcher herself):

HW: Back then . . . it (the giant panda) was . . . it was a real testament to our spirit.
YG: OK.
HW: If you ask me for more reasons, I really can't say. That's the reason I liked it.
YG: OK. So, what did you think of the giant panda when you first heard about it?
HW: It was tied to the admiration of Chairman Mao.
YG: Admiration?
HW: Exactly. Really! Our generation was like that.
YG: Hmm. I understand.
HW: Maybe you don't get it. Premier Zhou said, "When we send our beloved giant panda to someone, we receive something in return from our international friends". It was like that. For us, it was a kind of political reverence. Oh my!
YG: So, back then, how did you understand the cuteness of the giant panda?
HW: It was a national treasure at that time. Understanding . . . how do I put this into words for you? How do I express it? I would say it was something spiritual . . .
YG: OK.
HW: Nowadays, it's definitely appreciated for its beauty. But back then, the giant panda represented a kind of spiritual . . . a spiritual fulfillment of a dream.

YG: I see. That's quite interesting.
HW: Exactly. For others, they might come just to see the giant panda. For us, it was a spiritual restoration, or a spiritual sustenance, like a . . .
YG: Spiritual sustenance?
HW: Or . . . or . . . a kind of spiritual homecoming, that's the feeling we sought.
YG: Did you find it?
HW: Yes, I did. Oh, so that's what a live giant panda is like. Before, we only saw it in photos, just like that. Oh . . . back then, people's spirits were tied to the admiration of Chairman Mao. You might not feel this, but our generation did. I just remember the giant panda being a national treasure. Oh, and we sent it abroad. Wow! This is a giant panda. This is ours. It's our national pride.
YG: OK.
HW: It made us . . . it became a political star. We saw the giant panda in that way. When African mangoes were sent to Chairman Mao as a gift, he expressed gratitude to our African friends. Back then, those mangoes felt like sacred things.
YG: Haha.
HW: Oh! It was really like that, I'm telling you. You might not understand, but you can ask old men and women in their fifties or sixties; they'll say the same thing.
YG: It's a memory of that era.
HW: Yes, it truly is a memory of that era! But now, I support the idea of renting out the giant pandas instead of gifting them. Chairman Mao was too generous!

While Huang Wei hastened to justify his interest in the giant panda based on his admiration for Chairman Mao and Premier Zhou, emphasizing the generation gap between himself and me (and my research assistant, a recent university graduate in her early 20s), both my assistant and I nodded in understanding and approval of Huang's experience. We understood Huang Wei and his attachment to the giant panda and Mao's leadership well, as similar stories of admiration for Mao had been shared with us by our parents and grandparents over the years. In films, artworks, performances, and social media, the nationalist passion for Chairman Mao as the country's beloved figure has been a theme elaborated in various ways. In fact, we understood that there was no other way for Huang to love the giant panda in his generation except through his admiration for Mao. However, it is notable that Huang consciously justified his love for the giant panda by contextualizing the "time" and "generation" he lived in. He was aware that what seemed crucial and natural to him might be judged as ridiculous or irrational by younger researchers. But we did not share this view – ridiculing Huang's political attachment to the panda could also mean dismissing the experiences and stories passed down to us by our parents and grandparents about life in China and being Chinese over the past two centuries.

While Huang Wei's story suggested that the political enthusiasm for the newly established nation and its political leader certainly shaped people's interest in a nationally treasured animal, for the younger generation, Zhou Lan and He Dong's love for the giant panda had become more of a destiny. For Zhou and He, their affection for the animal did not stem from a gesture of friendship by Chairman Mao or Premier Zhou, but from the simple fact of living as Chinese people. Lin Qi suggested that the giant panda, as a China–West ambassador, is the story she would pass on to her grandson. The symbolic value of the giant panda served as intergenerational knowledge, connecting Chinese lives and reproducing meaning and consistency between the generations.

If we say that Huang's panda fandom was conditioned by his devotion to Chinese leadership, Zhou, He, and Lin portrayed a panda love that can be described as ideal and perfect: Love at first sight, unconditional, lasting, committed, honorable, and generous. In contrast, before the establishment of the People's Republic of China in 1949, the giant panda was primarily a visitor attraction appealing to the interests of Western audiences. How does an animal, once largely unknown to the Chinese public, rise to become an icon of universal love and devotion? What does this transformation reveal about the nation's relationship with the animal? If the stories of the participants show that this panda love always coincides with one's identification and experience of Chineseness, why does the nation-state necessitate love for an animal? Why, after all, is there such an interspecies love and affection? If panda love parallels one's identification with the nation, what does this love say about the nation?

In previous chapters, I have explained that the giant panda, initially sought after by Western nations, became the cause of suffering for local Chinese communities, the Chinese diaspora, and the nation's diplomatic strategies. In other words, since its discovery, the giant panda has symbolized the exploitation of colonial and imperialist powers over another culture, and the capitalization of natural resources at a relentless pace. In this light, the giant panda is tied to what the Chinese nation refers to as the "century of humiliation", contributing to the suffering and indignity experienced by the Chinese people under Western powers. Songster (2018) uses the term "panda nation" to suggest that the giant panda is a byproduct of the rise of the People's Republic of China, contrasting the animal's contemporary significance with its prior lack of established historical meaning. As I have discussed elsewhere, the giant panda's lack of presence in traditional culture and its insignificance to the Chinese before the establishment of the nation-state do not mean that the panda has no history. Instead, this history – the exploitation of the panda, the free labor contributed by the Chinese local community, the mourning for lost homes, and the co-presence in transnational cooperation – has often been bypassed or dismissed as Chinese cultural indifference, personal fantasy, or political maneuvering, terms that concepts like Orientalism and Othering have readily captured.

Interestingly, it seems that the tendency to overlook the panda's relationship to what can be termed suffering and exploitation is not solely for the benefit of

Western audiences. For the Chinese, the giant panda must be an animal stripped of its history in order to be adored. The real question regarding the panda's rise as a national icon and the Chinese fandom surrounding it is not how the panda rose alongside the nation's growing global influence, but how it transformed from what could be considered a "mixed blessing" and a "tasteless necessity" into a beloved symbol of authentic, lasting, unique, and devoted love from people who identify with being Chinese.

"Only China has pandas"

One of the most frequently mentioned reasons by participants for their conversion into panda fans was China's unique ownership of the species. For example, Zhu Xi (F, 55), a dedicated fan, suggested that pandas "are born with the mission of showing the world that there is such a unique and magical species on Earth, one that only exists in our country". Huang Hua (F, 34, junior fan) noted, "I think pandas are very special to our country because they are rare animals that only China has". She further emphasized China's ownership of the animal: "Every single one (panda) should be brought back to our country. Let them (foreigners) come and see, right? If they . . . if they really like it, they should come all the way here themselves. Don't . . . don't lend the pandas out". In Huang Wei's earlier discussion, he also mentioned that he preferred renting the giant panda and that Chairman Mao was too generous when he gifted pandas to other nations. Similar claims regarding China's unique ownership of pandas were echoed by other participants, with a few considering this claim of ownership the primary reason for their interest in the animal.

This collective awareness of the panda's ownership, declared through the lens of the nation-state, forms the core of the panda fandom. What is particularly fascinating about this panda fandom is how it reinforces the giant panda's stately private possession – a feature emblematic of capitalist relationships – through individual attachment and affection toward the nation-state. This ownership of the giant panda provides a sense of pride and achievement, which in turn strengthens the individual's connection to the nation. The attachment to the panda fandom conveys a sense of belonging, meaning fans are granted free space and flexibility to experience the giant panda. Huang Hua's statement exemplifies her attempt to leverage the cultural and geographical differences between China and non-Chinese people – "let them (non-Chinese) come and see the giant panda if they liked it so much". While the Chinese nation-state privatizes the ownership of the giant panda, it seems that those who identify with Chineseness become equal stakeholders in sharing the panda as private property. Identifying with Chineseness thus creates a unique relationship to the animal – one where the species is under the control and manipulation of anyone who owns and identifies with Chinese culture.

Despite the giant panda's lack of traditional symbolism in Chinese culture and its more recent exploitation and humiliation associated with the Chinese community, what grounds the panda in her Chineseness is the transformation of the animal into a nation's private property. In other words, the Chinese community's established claim to the giant panda does not depend on the elaboration or re-establishment of certain Chinese traditions, but rather on the capitalist relationship introduced first by Western nations and cultures. The celebration of China's ownership of the giant panda, viewed through this lens, foregrounds an already "Westernized" Chinese community (Chow, 2003), whose relationship with the species is channeled through the notion of private ownership. The affirmation of this ownership mediates the community's experience as a unified whole. The statement "Only China has giant pandas", at the heart of the panda fandom, should be understood as an alignment and recognition of the necessity and significance of private ownership in capitalism – essentially asserting that China, as a nation and a community of people tied to her, can be listed and identified as a participant in the ongoing capitalist relationships between the West and the rest of the world.

Moving away from the giant panda diplomacy, which Huang Wei interpreted as Chairman Mao's generosity, Huang also supported the giant panda rental strategy. His opinion, coming from a 65-year-old Chinese person, affirms the Chinese public's transition into a capitalist relationship with the West. Nicholls (2012) notes that the United States had refused to accept the entry of panda Chi-Chi because she was "a communist good" during the panda diplomacy era. The label "communist" might obscure the panda's attachment to Chineseness and its political connotations. However, the truth embedded in the term is that the giant panda, fundamentally, was a "good" – a unique and ubiquitous presence in capitalist relationships, whether in China or the United States. This suggests that beneath the nation's connection and appeal to the giant panda, the connection affirms the capitalist agenda and mode of production. When the French priest David and the American hunter Harkness transported the giant panda beyond China's borders in the early days of the panda fandom, the Chinese community remained largely outside the capitalist relations and was unable to navigate the emerging task of capitalization. The local landlord Wang, the Chinese locals, and the diaspora Jiang were confused and frustrated in their attempts to understand and connect with the animal. In contrast, the statement "Only China has giant pandas" suggests a sufficiency: On the one hand, a community/nation named "China" could mediate the Chinese people's relationship with the animal; on the other hand, the panda is exclusively owned by this community, grounded in capitalist principles. The state and capital provide the apparatus and means of production to materialize the Chinese people's connection to the animal.

The appeal of the panda fandom, therefore, lies in its embrace of both the capitalist agenda and the nation-state as a trope supporting the capitalist mode of production. The construction of a "panda nation" (Songster, 2018) mandates a path that is underscored by both the ideology of a specific, local understanding of the nation and participation in a broader, globalizing capitalist movement. The panda fandom achieves the dual task of bonding people through the idea of a shared nation while simultaneously exploiting the labor force of both the people and the giant panda as a resource, all mediated by the notion of the nation-state.

Stokes (1986), in addressing the question "how is nationalism related to capitalism?", argues that the concept of the nation and its nationalist concerns – often irrational and narrow in focus – seem, at first glance, to be in conflict with the broader political organizations implied by the international division of labor. According to Stokes, the capitalist mode of production requires mobile, literate labor forces whose focus has shifted from traditional kinship ties in families, clans, or villages to more professional and complex roles facilitated by larger social institutions like schools or the nation-state. Gellner (2008), in *Nation and Nationalism*, defines nationalism as a political principle that strives to align the political unit with the national unit, an alignment that became inevitable with the advent of industrialism in an ethnically differentiated world. Gellner posits that nationalism is not only a political ideology but also a necessity for the functioning of modern industrial societies, where educated, mobile, and homogeneous workforces are created and integrated into the capitalist mode of production. Thus, Gellner considers the nation and nationalism to be modern phenomena inherently tied to the structure and mechanisms of industrial society. Smith (2013) further develops Gellner's ideas by observing the persistent and intensifying nationalism in contemporary society. He shows that while globalization has prompted unprecedented interconnectedness, technological advancements, and the spread of global culture, the construction of the nation-state has been crucial in cultivating a sense of national belonging and distinct cultural uniqueness among its citizens.

In a more recent study, Cox (2020) agrees with Gellner's view, suggesting that the capitalist economic roots of nationalism and the formation of the nation-state can help explain the co-occurrence of economic inequality and growing ethnic diversity in many parts of the West. Cox concludes that nation and nationalism create collective emotions that connect to the instrumental bases of social and political everyday life. As Gellner observes, in pre-industrial societies, horizontal differentiation based on economic and social divisions was embraced through local and family-bound stability. However, in industrial societies, there is a shift toward strengthening the boundaries between nations rather than between classes. This shift aims to tap into deeper emotional and psychological needs for belonging and unity, while simultaneously enabling the production and management of a population as the needed labor force.

In his conceptualization of biopower – a form of power targeting specific populations to optimize their health, life, and productivity – Foucault (1990) aligns with the view that the nation-state becomes the necessary apparatus to accommodate and negotiate the complex, often conflicting dynamics within populations. Foucault (1990, p. 141) explains the role of the nation-state apparatus in this way:

> If the development of the great instruments of the state, as institutions of power, ensured the maintenance of production relations, the rudiments of anatomo- and bio-politics, created in the eighteenth century as techniques of power present at every level of the social body and utilized by very diverse institutions (the

> family and the army, schools and the police, individual medicine and the administration of collective bodies), operated in the sphere of economic processes, their development, and the forces working to sustain them. They also acted as factors of segregation and social hierarchization, exerting their influence on the respective forces of both these movements, guaranteeing relations of domination and effects of hegemony. The adjustment of the accumulation of men to that of capital, the joining of the growth of human groups to the expansion of productive forces and the differential allocation of profit, were made possible in part by the exercise of bio-power in its many forms and modes of application. The investment of the body, its valorization, and the distributive management of its forces were at the time indispensable.

According to Foucault (1990), biopower and its application rely on the development and construction of the nation-state as the primary apparatus for managing, manipulating, and operating through a population. The nation-state apparatus anticipates and designs the life of a population, with its biological sphere and bodies becoming the primary focus of education and investment. For Foucault, the construction of the state apparatus does not eliminate class differences, cultural segregation, or social hierarchies. Instead, these differences – the effects of hegemony and domination – are reproduced within the structure and notion of the state. The primacy and centrality of the state apparatus overshadow the class conflicts that the capitalist regime inevitably nurtures, while a collective and shared community, facilitated through the apparatus of the state, emerges.

The statement "Only China has pandas" creates a community centered around the image of the giant panda. The parallel between the giant panda and the nation of China within this community establishes an equation in which one's love for the animal signifies one's submission to the nation-state, thereby fully participating in the nation's management of its productive forces. In this context, the panda fandom serves as a national appeal for the community and is universal for the nation's population. To understand this dynamic further, we need to explore the productivity and modes of production that the panda fandom, or the central appeal of the panda fandom, unfolds and maintains through the apparatus of the nation-state.

The ideal love

Liang Mei, a 30-year-old female panda fan, visited the Panda Base for the first time with the task of introducing her newborn to the animal. Liang became interested in pandas when she saw social media posts about the return of a giant panda to China. Her national sentiment was deeply stirred when she learned on social media that giant pandas were being mistreated abroad. "They said the giant pandas raised abroad were very thin", Liang explained. "So, I paid extra attention to giant pandas on social media". As her interest in the giant panda grew, she decided to bring her child to the Panda Base.

> "I decided to bring my child here to see the pandas in person and let him experience it for himself. Because the giant panda is a national treasure. Once you have kids, or even for yourself, after you start working and have some financial stability, you want to learn more about it, right? Kids usually like animals, so they probably love the pandas. For us, the adult, it is also about a sense of national pride".

When asked if she needed to cultivate the national pride in her children, Liang said: "Anyway . . . we haven't really tried to indoctrinate him. Kids naturally pick things up. From kindergarten to elementary school, they gradually learn. Now, bringing him here is about letting him experience it. First and foremost, it's for fun, and then to teach him some basic manners. Also, we'll explain why pandas are so popular".

In our further conversation, Liang shared that her career as a primary school teacher helped her understand how children learn about the giant panda. She provided the following insights on introducing the giant panda to her classes:

> Sometimes, for example, during our class meetings, we might educate the students about these things because it's basic knowledge that everyone should have. Our class meetings follow a structured system, which includes specific lessons on national treasures, the national flag, the national emblem, and similar topics. We make sure to cover all of these with the students.

Liang Mei's perspective, as a primary school teacher, offers valuable insight into the reflections of Zhou Lan and He Dong, who explained that their initial encounter with the giant panda laid the foundation for their enduring affection for the animal. According to Liang, the image of the giant panda, along with national symbols such as the national flag, is systematically taught to children because it is considered "basic knowledge" and "common sense". Liang Mei felt that it was unnecessary to indoctrinate her own child because children naturally gravitate toward animals. Her child would develop his own love for pandas based on the education he received and the care he experienced from his parents. It is easy to imagine that when Liang Mei's child grows up, he will recount his encounter with and love for the giant panda in much the same way Zhou Lan and He Dong have shared their stories.

In an empirical study of participants meeting potential partners for the first time, Zsok et al. (2017, p. 7) argue that love at first sight may not involve intense passion or immediate feelings of love, but rather "a readiness to experience them at best". The researchers suggest that love at first sight can signal a strong initial attraction that sets the stage for deeper emotions and affection to develop. Constructing this love at first sight, either retrospectively or in the moment, plays a key role in how individuals interpret and experience meaning in the relationship. Building on the ideas of Zsok et al., we can say that Zhou Lan, He Dong, and Liang Mei's child were "ready" to love the giant panda from the moment of their first encounter. In fact, one could argue that the depth, authenticity, and lasting nature of their bond with the animal had already been "readied" from that very moment.

Cotter (2016), citing the work of Alexandra Kollontai, suggests that love is a "profoundly social emotion" that circulates through ideological and illusory articulations. Cotter (2016) further elaborates:

> "Love" is the subject of social contestation because of its historical role in strengthening the bonds of social relations which have been made historically and materially necessary to a given mode of production. Love is not transhistorical produced under definite conditions. Specific forms of love are produced and become culturally "valuable" within a given set of material relations of necessity for the way in which they help to reproduce the existing social relations of production. In class society, this means that the dominant forms of love become dominant or culturally valuable because they help to maintain the existing class relations and to reproduce the conditions necessary for maintaining class relations.

The panda fandom, building on the panda–human love circulating between the animal and the Chinese community, is one such socially produced emotion that is anchored in a set of material relations. This point will be revisited in the next section. For now, it is important to understand that the universal form of panda love across the nation of China, along with its articulated social expression embraced by every Chinese, should not be seen as an emotion that is purely authentic, bodily, or natural. Instead, as I have demonstrated, the state apparatus serves as the primary vehicle through which panda love as a specific emotion and affection is shaped and conveyed between the Chinese community and the animal. The Chinese love of the giant panda was not an accident – a deep and authentic romance that randomly falls upon its members. Rather, panda love is a calculated and deliberately cultivated experience lived and experienced through bodies marked as Chinese.

Although panda love may seem like an idealized experience, resonating with the romantic notion of love since the modern era in both its depth and strength, what distinguishes panda love from personal and individual romantic experiences is its universal impact on populations and its emphasis on intergenerational heritage. The three quotes below from panda fans illustrate how the giant panda – its symbolic significance and conservation efforts – has come to be viewed as a significant cultural heritage shared between generations of Chinese. In his reflection on the making of heritage in the United Kingdom, Hall (1999, p. 5) states that the heritage of a nation "constructs identities by selectively binding their chosen high points and memorable achievements into an unfolding 'national story'". Hall (1999, p. 6) argues that a national heritage is selectively constructed to imagine a society that is "culturally homogeneous and unified". Because of its potential to bind each member individually into the larger national narrative, Hall (1999) suggests that national heritage can be a powerful source of meaning and a point of identification.

In this context, Hall's (1999) articulation of national heritage, based on the concept of "Englishness" and its virtues, was aimed at distilling the centrality of Western hegemony in order to create space for minority relationships, represented by the "otherness" of non-Western ethnic minorities in the United Kingdom.

> I think it means that . . . it should be . . . how should I put it? It's an honor for us to see them, first of all. Secondly, the growth process of pandas is a learning experience for us as a nation, teaching us about nurturing. This might influence the education of the next generation, instilling certain values step by step. After all, pandas are a national treasure. As citizens, we should all have a basic understanding of this, right? These are just the thoughts of an ordinary person. The giant panda is a cultural heritage.
>
> *(Huang Hua, F, 34, junior fan)*

> I think giant pandas are indeed national treasures that everyone should protect. Over the past few decades, our country's economic development has also emphasized the importance of wildlife conservation. This focus on protecting wild animals will continue to grow more important with each generation.
>
> *(Wu Ying, F, 65, junior fan)*

> Moreover, since I was very young, pandas have been endangered animals. Now they are classified as vulnerable. This change reflects the efforts of one or two generations in protecting the natural environment. Turning an animal that is very important to China into a symbol of national pride and cultural significance.
>
> *(Hou Yi, F, 42, junior fan)*

While probing the specific "Chineseness" that the giant panda, as a national heritage, embodies is an important topic, this extensive discussion goes beyond the current scope of this book. Here, it is sufficient to say that panda love has been constructed as a national heritage, with an influence that extends far beyond the immediate regulation or discipline of a single generation. Rather, the particular biopower that panda love organizes targets generations – a lineage view of populations across time. The future Chinese generation depicted in the panda fandom reflects a dynamic of learning, built upon knowledge acquired about the giant panda, and a promising future that embraces better environmental and conservation practices. This vision of the panda fandom anticipates an improved Chinese lineage through generational productivity, economic development, and environmental protection.

Anderson (2012, p. 29) notes that the affective life of individuals and collectives has become an "object-target of" and "condition for" contemporary forms of biopower. The panda fandom projects the affective life of the Chinese community through the experience of panda love. In its growing influence and impact, the panda fandom affirms the Chinese community's capacity to affect and be affected through a heightened sense of panda–Chinese–nationhood infatuation: To love one's nation and national treasure unconditionally, lastingly, readily, timelessly, and passionately. The young children of the nation are invested in and educated as capable bodies who, upon seeing the giant panda, are bound to love the animal. In fact, a closer examination of the narratives from Liang Mei, Zhou Lan, Huang Hua, Huang Wei, and He Dong reveals that the opposite – such as not loving the giant panda – has not been considered an option. As Chinese, the possibility of imagining or envisioning a different relationship with the giant panda simply does not exist. In contrast to this contemporary, passionate love, we see the landlord Wang and local communities' tolerance of Roosevelts' hunting of giant pandas in the 1920s, and Jiang Yee's ambiguous and melancholy fantasies about the giant panda and its fate in the London Zoo. Andersen (2012, p. 36) concurs with Cotter in stating that "the market is simultaneously the limit of and site of verification for government action", showing that biopower governs and affects public life according to the principles and valuations of the market, which forms the cornerstone of existing social relations of production. The key question that the panda fandom must address is the particular social relations central to the current mode of production that this panda–Chinese love sustains and nurtures.

People surplus

> I think . . . well, it's our national treasure, right? It's a special product from Sichuan, and it's quite rare. It's an endangered species. This is something unique to Sichuan. As a Sichuan native, I'm especially proud. It's also considered a mascot of good luck. Just look at all these people, the crowds coming here. They're not just here to admire it; there's a sense of care and love too. We hope our pandas will continue to thrive and flourish, always remaining our national treasure.
>
> Pandas don't cause trouble; they don't provoke conflicts. They are symbols of peace, much like doves. Having them makes foreigners envious. They are not just valuable; they are beyond priceless, more noble and dignified than any material wealth. Foreigners admire our pandas. They are rare, gentle, and lovable, causing no harm to people or livestock. They are harmless animals that everyone cherishes.
>
> (Chen Jie, M, 65, junior fan)

Chen Jie visited the Panda Base with the intention of observing people – the crowds flowing around him, which brought him a sense of happiness and satisfaction. In his description, Chen depicted two life forms. On the one hand, there was the rare and endangered species that was uniquely and exclusively owned by Sichuan, China. On the other hand, there was the human crowd, symbolizing the thriving future of the nation. While Chen suggested that he focused on the crowds and their movements, it is the intersection of these two life forms – the human surplus and the panda's primitive existence –that shaped his experience at the Panda Base. The question arises: How could these two seemingly contrasting life experiences – the lives of the Chinese people and the panda – intertwine to create an image of thriving and flourishing?

If the nation of China claims ownership of the giant panda, as suggested by Chen, then the nation's prosperity and thriving also draw on its "ownership" of the Chinese population. The crossing of the panda's life and the Chinese people's life in the imagined space of the nation shows that panda love is not only affection for the animal but also a bond, a claim of possession of a population whose lives and forces continue to contribute to the nation's development. As Hou Yi's quote suggested, the efforts of one or two generations have been devoted to the conservation of the natural environment, and future generations will continue this work. Chen's observation – that the Chinese population is flourishing – aligns with Hou's vision of a future where more generations commit to national projects such as conservation and panda protection. The nation's power to manage, regulate, and exploit both the labor of the rare animal and the Chinese population contributes to the nation's future development and economic growth.

The parallel between the Chinese population, as a crowd, and the giant panda, as an individual, raises further questions: If Chen believed that the giant panda's value was beyond priceless, how should we view the Chinese population surrounding the rare and endangered animal? Is the Chinese population equally priceless? If the giant panda is admired by foreign audiences, does the Chinese population experience the same level of admiration? Does the value of the Chinese population receive the same recognition as the panda's?

The trick, however, is that Chen's comments did not suggest an equal valuation between the panda and the Chinese population. The giant panda's "beyond pricelessness" was not determined by the Chinese population, who were present in the same place and context as the animal. Instead, it was foreign audiences, particularly the Western gaze, that served as the invisible yardstick by which the value and significance of the giant panda were measured. In this way, the question arises: What role does the Chinese population play in this scenario, as they swarm around the giant panda?

A dedicated panda fan, Tang Xue's (F, 31) comments further help explain the Western gaze:

TX: Hmm. There are also some prejudices, right? Some people think, oh, what's so special about pandas? If you're in the right circles, it's fine. But in general society, people might think, oh, all pandas look the same. First of all, it's a national treasure, right?

YG: Right.

TX: It's a treasure. What does it mean to be a national treasure? It means it's something precious.

YG: OK.

TX: It should be cherished; it should be respected. Sometimes, I even think that we, as humans, might not be as precious as these pandas. I might think that way. They really like Hua Hua, I think that panda . . .
YG: Has completely crossed borders?
TX: Yes, yes. And pandas like Hua Hua and Fu Bao are quite famous.
YG: OK.
TX: And this fame might give the international community a good impression of China.
YG: A good impression?
TX: Yes. It's like a bridge, in a way.

While Chen Jie groups the Chinese population and the singularity of the giant panda in a shared space, Tang Xue explicitly suggests that the value of a giant panda can sometimes exceed that of a human being. One might argue that Chen's statement implicitly suggests that an individual Chinese life is not comparable to the life of a giant panda. Both fans, however, share the belief that the giant panda is not only a symbol of national pride but also an entity that connects to the West – where the value of the giant panda is often determined, and where the impression of China is influenced. Even though the interviews took place between Chen Jie, Tang Xue, and myself, the giant panda seems to merge with a Western presence and audience, consistently shaping the standards and expectations for what it means to be Chinese.

The notion of the giant panda being "beyond priceless" in the eyes of both the West and the Chinese population, whose value remains largely undetermined, can be contrasted meaningfully with Harney's (2008) observation in *The China Price*:

> Wages in China are extremely low compared to those in developed countries. On average, Chinese manufacturing workers are estimated to have earned $0.57 an hour in 2002, which translates to about 3 percent of the average hourly compensation of manufacturing employees in the United States and other developed countries. Chinese labor costs are a fraction of those in Mexico, Brazil and developed Asian countries. By one estimate, they are even lower than wages handloom operators earned in the early Industrial Revolution in the UK or workers in a Chicago lumber yard took home in the mid-nineteenth century.

Chuang (2020, p. 6) in a recent study of China's low production advantage resonated with Harney's observation:

> Universal land-use rights in the countryside allow individual laborers to work for employers in cities while leaving their families behind in villages to farm land to defray the cost of food and supplement low wages. Since workers can retreat to farming during periods of unemployment, they can withstand irregular labor contracts and low wages. This has made China's labor force appear

self-replenishing, peopled by migrants from impoverished inland regions who cycle in and out of work while an ongoing subsistence farming economy subsidizes wages and absorbs unemployment.

When I inputted a command to ChatGPT 4.0, requesting an image of a giant panda amid a crowd of people, I believe the artificial intelligence effectively dramatized Chen Jie's experience of China's prosperity at the Panda Base. The simplicity of the image helps clarify the concept of population surplus, where the value of the population is juxtaposed with the image of the giant panda. In this image, the giant panda (represented by the dark, round bear head) occupies the center, while the human crowd (depicted by blue dots) is spread evenly across the scene. However, it is important not to equate the value of the giant panda – its "beyond pricelessness" – with its central, enlarged position in the image, which makes it appear larger and more prominent than the human figures.

In fact, the human population also occupies central space, extending across the image in a way that the giant panda is refrained from. The population surplus, I believe, comes from the repetitiveness and similarity shared by each human figure. It represents a population, rather than a collection of unique individuals. The crude manner in which the artificial intelligence sizes the image – where human figures are cut in half at the edges in a careless and random way – suggests how individuals are treated as part of a population, disposable and interchangeable. It is easy to imagine that the human population extends beyond the frame of the picture, and the half-human figures in this image could be quickly replaced or substituted based on the size of the population.

If the giant panda expresses its pricelessness and value through its singularity and integrity, the value of the Chinese population can only be understood through its perceived cheapness, embodied in the homogeneity and similarity of its individuals. These cheap individuals can be supplemented, replaced, cut, disposed of, removed, and reduced precisely because of their inclusion within the larger population. The value of the Chinese population, as Harney (2008) and Chuang (2020) argue, is tied to the living labor bound up with low wages and produced in vast, unimaginable quantities. This can be understood in terms of what Heinrich (2018, p. 5) describes as the "global corporeal surplus". Huang Hua, Wu Yin, and Hou Yi's comments further suggest that the valuation of the Chinese population extends not only to the present but also across past and future generations, spanning across time.

What is more unsettling is not merely how the value of the Chinese population is shaped by its connection to the priceless and singular giant panda, but how a persistent Western gaze has always been present, legitimizing and measuring the values of both the panda and the Chinese population. If we highlight that both Tang Xue and Chen Jie invoked the admiration and impressions of foreigners to validate the value of the giant panda, it is not an exaggeration to argue that this foreign gaze also determines the value of the Chinese population and how this population perceives its own worth. While panda fans portray the giant panda as a bridge connecting China and the West, this consistent, though often invisible, presence of the West through the image of the giant panda suggests that the panda fandom, though seemingly a Chinese national appeal, has also been an ongoing process of Westernization and industrialization. When Foucault (1977b) repeatedly emphasizes that power is omnipresent, not concentrated in a specific institution or class, but permeating every aspect of life, the words of panda fans testify to how Western hegemony, the modernist agenda, and the capitalist regime have become routine aspects of Chinese life to the point that fans such as Chen Jie have come to admire and appreciate the structure and mechanism of this power display.

Perhaps it is not the invisible and persistent presence of the West in the capitalization and valuation of the giant panda that is most disturbing, but rather how panda fans – the Chinese population itself – have come to accept the violence and exploitation embodied in the valuation of both the Chinese population and the giant panda. This acceptance, expressed by Chen Jie, a 65-year-old man who has experienced early Chinese modernization in the 1970s and many of its traumatizing consequences, reveals a continuity between two seemingly distant historical periods. What is captured in the panda fandom, in Martin Heidegger's (1977) terms, are the bodies of the giant panda and the Chinese population, arranged, cropped, and disciplined through a specific visual technique, while a Western subject and audience coexist with the image. Chen Jie's pleasure in seeing the prosperity of the panda fandom suggests that he, too, holds the privilege of viewing from the perspective historically occupied by the Western audience – despite this Western figure remaining central to the visualization and objectification of the Chinese population and their bodies.

But how could Chen Jie, an ordinary citizen and panda fan of the Chinese nation, be granted or entitled to such a privilege – the privilege of seeing from the subjectivity that has traditionally been the domain of the Western audience? Could Chen, someone whose fate is tied to the Chinese population, escape the role of being a part of the world picture and instead become a viewer of it? Was it sheer luck? If Chen could, somehow, shift from being part of the image to taking on the role of the observer, does this mean the dominance of the viewer and the vulnerability of the gaze have been reconciled by his newfound freedom to move?

The overlap between the pleasure Chen experiences from viewing both the Chinese population and the giant panda, alongside the authentic and lasting love the Chinese population devotes to the nation-state, offers a clue to this visual dialectic: The nation-state of China is the precondition that enables Chen to adopt the role of a viewer within the world picture. In other words, Chen, as an individual of the Chinese population, does not access the experience of viewing through his individuality or personal body. Rather, to observe the panda and the Chinese population from a distance requires the social apparatus of the nation-state, which creates the vantage point for the Chinese population, like their Western counterparts, to grasp the visualized world. This relationship between panda fans and the nation, intertwined with the pleasure of viewing, reinforces the community's attachment and commitment to the nation. The panda fandom builds on an economy that circulates the living labor of both the Chinese population and the giant panda, with the nation-state apparatus supporting the valorization of this labor within the capitalist world. The love for the panda compensates for the exploitation and suffering that inevitably accompany the management and manipulation of populations. In the next section, I will explore what this economy of the panda fandom means for the giant panda herself.

Panda primitivity

The seemingly absent role of the giant panda, despite the Chinese love for the species, the Chinese nation, and the exploitation of the Chinese population being key cornerstones of the panda fandom, is not, interestingly, due to the panda's lack of significance, but rather because of her reality. How could an animal that occupies the center of the panda fandom somehow be at the margins of the social relationships on which the fandom is built? Why should we question the love for the panda when a nation's population has wholeheartedly and devotedly embraced the species? In fact, I would argue that it is precisely through this fated, love-at-first-sight connection between the Chinese and the giant panda that we must question this tightly intertwined multispecies entanglement. However, to understand this panda love and its effect on the animal, we must examine the social and material relations that this love strengthens and what they mean for the panda.

The question that first drew my attention to the panda fandom was how this strongly experienced and passionately articulated love affair with the giant panda resonates with the concept of multispecies existence (Kirksey & Helmreich, 2010; Pacini-Ketchabaw et al., 2016; Van Dooren & Rose, 2012), the more-than-human world (Abram, 2012; O'Gorman & Gaynor, 2020; Tsing, 2013), biodiverse environments (Haahtela, 2019; Lovell et al., 2014; Schebella et al., 2020), the Anthropocene (Malhi, 2017; Steffen et al., 2011), natureculture (Haraway, 2013; Latimer & Miele, 2013), and the post-human (Braidotti, 2013; Schussler, 2022) – all of which have become significant concepts in conceiving and imagining life with nonhuman others. My intuition was that because the panda fandom represents a place where the public's affection and adoration for the giant panda have reached an unprecedented intensity, exploring the human–panda bond should remain the central focus of any study of panda fandom. This is the task I address in Chapter 9, where I will delve deeper into the human–panda relationship and how it can potentially move beyond the existing social and material contexts. This section, meanwhile, offers a closer examination of the giant panda's consistent role as a means of production in the panda fandom, an observation that has been discussed in previous chapters. My thesis is that the panda fandom creates social conditions in which the conflicts between workers and the means of production are reconciled.

From Chapters 1 to 7, I have demonstrated that the social and material relations that give birth to the panda fandom are rooted in the Chinese community's ongoing participation in transnational capitalist transactions with Western nations. The panda fandom began with a Chinese community seeking trade, identity, and survival while confronting the violence of imperialism, colonialism, and capitalism. Today, the panda fandom continues to operate within the existing capitalist regime, with biopower controlled and managed by apparatuses such as nation-states and the trope of national love. Emphasizing the Westernization of the Chinese community within the panda fandom shows that little, if any, fundamental change has occurred in the way the Chinese community interacts with the giant panda – especially at the early stages of the giant panda industry and fandom, when the survival of the Chinese community was a matter of life or death. This does not only mean that the Chinese community has inherited Western capitalism's exploitation of natural resources, such as the giant panda, but also that the Chinese community has become dependent on the oppression and super-exploitative use of natural resources for its urgent need to become one of the world's more developed economies. There remains little space for the Chinese to critically reflect on their established relationships with nature, animals, and other resources.

As Chapter 4 discusses, panda diplomacy has been the Chinese nation-state's deliberate move to capitalize on the value of the giant panda to engage in trade with Western markets. Despite this materialization of the giant panda being seen as the Chinese state's politicizing strategy, manipulated by the West (Hartig, 2013), China's diplomatic use of the giant panda consolidates the nation's private ownership of the species. In other words, panda diplomacy allows the Chinese nation-state to politicize a social–material relationship rooted in the capitalist exploitation of the giant panda. For the animal, this West–China exchange confirms her status as a means of production – a super-exploited natural resource linked to human society through the means of private ownership. As the panda fans I quoted in this chapter have illustrated, the nation's ownership of the animal is both the source of the fans' national pride and the primary link between the fans and the animal. Despite its struggle and resistance against Western influence and hegemony during its early stage of modernization, the Chinese community did not become "enlightened" by choosing to revolutionize the capitalist relationship with the animal and emancipate the animals as an oppressed class. Instead, like its Western counterparts, the Chinese community found in these underprivileged animals a source of value and utility that helped reconstruct, renew, and rejuvenate the Chinese community's cultural and material production, both in substance and form.

The panda fandom is a social sphere where the super-exploitation of the giant panda is reproduced and reinforced through the mediation of the nation-state apparatus. Chow (1995, p. 22) suggests that capitalism maintains its mode of production through intellectual elites' interest in exploring the "primitive passion", typically associated with "the animal, the savage, the countryside, the indigenous, the people, and so forth, which stand in for that 'original' something that has been lost". It is not far-fetched to suggest that capitalism, in its materialist mode of production, replicates the "primitivity" of natural resources that, in turn, sustains the capitalist exploitation of nature – precisely because nature is contextualized as exotic, different, and external to human society, thus justifying its exploitation. The Chinese community and the panda fandom have sustained this primitivization of the giant panda rather than challenging the animal's oppressed state.

What also draws me to figure (the ChatGPT's image of giant panda) is the artificial intelligence's ability to parallel two seemingly opposite experiences in a crude and abstract way: The pricelessness of the giant panda at the center and the cheapness of the human individuals scattered across the image. The visual image affords a straightforward confrontation of these two extremes configured within the capitalist relationship. On the one hand, the giant panda, as the means of production, is privileged to the extent that both its use and exchange values have been rendered immeasurable. Thus, the panda's pricelessness is symbolized by its central placement in the image, a phantasmic image overshadowing a fraction of the human population. On the other hand, the human population exemplifies the extremity of cheapness – human figures in the image lack shape or individuality and are depicted as rows and columns of repeated figures, with little regard for their completeness. The irony of Figure 1, I believe, lies not in the contrast between pricelessness and pennilessness but in how the invaluable is congruent with the worthless. The juxtaposition of the panda and the Chinese population sharing the same visual space underscores their underlying connection rather than contradiction. It seems that the giant panda's value cannot be easily distinguished from the cheap living labor contributed by the Chinese population. What makes this juxtaposition of the giant panda and the Chinese population possible, if not their market values – which are inherently mismatched – is their shared experience of exploitation under capitalism. This "cheapness" of the giant panda, arising from the appropriation of the animal's labor, free product, body, and life cycle, marks the animal's disposable use, integrated freely into the capitalist production process.

When asked whether she would want to "be a giant panda", Lu Qian, a dedicated fan aged 36, responded as follows:

YG: OK. Um . . . so do you ever feel . . . hopeful about being a panda?

LQ: Hopeful, even just a little bit? Um, no, I don't really hope to be a panda.

YG: But . . . why do you think you don't?

LQ: Because, even though it's a bear . . . a panda; look, everyday it's like . . . they say, "Ah, I really want to be a panda! I can be idle all day". But the panda being idle is also within our control. What the panda eats, where the panda lives; it's all our decision.

YG: Hmm.

LQ: So, in my opinion, I don't really want to become a panda. I'd rather, you know . . . have my own life. Even though now I feel many things in my life are out of my control. But, at least a part of it, right? I can decide for myself on this.

YG: OK.

LQ: It is life and destiny. But I feel like pandas, it is just they are completely controlled by humans, aren't they?

YG: Yes.

LQ: So I don't really want to become a panda.

This refusal to live the life of a panda followed Lu Qian's claim that she considered the panda Hua Hua her "daughter" and sometimes felt that her concern for the giant panda was too obsessive. When I suggested that Lu Qian imagine her life as a giant panda – loved, treasured, and spoiled by many fans – her refusal was typical, as she was reluctant to trade her mobility and freedom for the confined and manipulated existence of a giant panda. Beneath her motherly adoration and love for the giant panda, however, is her recognition of the deprivation and oppression experienced by the giant panda through this very love.

Marx (2000) posits that the relationship between the working class and the means of production in capitalism is inherently antagonistic because capitalists own and control the means of production through private ownership, while the working class must sell their labor to the bourgeoisie to survive. Marx's theory would suggest that the panda fan community, like the working class selling its labor, should foster a conflictual and alienated relationship with the giant panda, which represents the means of production. However, in panda fandom, we see a love for the panda that almost seems to idealize every encounter between a Chinese individual and the animal. Does the panda fandom suggest an alternative social–material relationship where the conflict between living labor and dead labor in the capitalist mode of production has been reconciled?

In fact, in panda fandom, the conflict between private ownership and the exploitation of the working class appears to be marginal, as the class struggle and conflict Marx observed in capitalism are minimized by the appeal of the nation. The Chinese community's identification with the nation-state – "only China has pandas" – and the escalating passion and devotion to the unified ideology of a nation, perpetually influenced by the Western gaze, provide a different lens through which to understand the panda fandom. In the next chapter, I will explore the cuteness of the giant panda, another biopoliticized form of power that aestheticizes the animal–human social relationship, based on reflections from the fieldwork.

8

THE CUTE EXISTENCE

"A fourth kind of feeling"

A study of panda fandom is incomplete without recognizing the role of cuteness. Chapter 3 theorized cuteness as a form of biopower that disciplines bodies, regulates other lives through the human image, and capitalizes on surplus profit generated by those bodies. In this chapter, I further explore the aesthetic of cuteness in the giant panda fandom, focusing on fans' daily engagement with the biopower of cuteness through the animal's super-exploitation. Specifically, I investigate the panda's cuteness through the fans' entanglement with it – how Scott's cost-saving animal becomes a quotidian reality for panda fans and why cuteness reflects the capitalist ethos of efficiency, cost-reduction, and affordability for fans. Meanwhile, the animal experiences exploitation and estrangement from her animal life within the panda fandom.

Ngai (2010, p. 12) highlights a utopian dimension of cuteness, suggesting that encounters with the cute involve "a desire to inhabit a concrete, qualitative world of use as opposed to one of abstract exchange". Ngai notes that Marx (2000) situates this desire for intimacy and physical entanglement within the realm of consumption, meaning that engagement with the cute explores the commodity's use value. Given that the giant panda is a means of production – privately owned property rather than a commodity – I demonstrate that panda fans do not merely consume the giant panda. Instead, they cultivate proximity and connection to the animal as part of "making life", a homogenized labor governed and dictated by Foucault's concept of biopower operating through the population.

Cuteness as biopower

In Chapter 3, I explained how cuteness, as a form of biopower, targets regulating and managing the human population through its life-enabling origin and its power to make life. I suggested that cuteness's scientific basis lies in Darwin's

DOI: 10.4324/9781003511229-9

theory of evolution, where its generative connection supports human infants' survival through adult care. The biopower of cuteness aestheticizes this evolutionary instinct, transforming biological knowledge into a materialized fact embedded in the universality of capitalist relations. As capitalism holds a ubiquitous, global presence in human societies, it also serves as a platform where biopower generates scientific knowledge, truth, and social norms, such as the evolutionary mechanisms underlying cuteness.

As biopower, cuteness leverages the natural affection between human adults and infants, reproducing this bond within capitalist relationships. This reproduction of generative instincts within the sphere of consumption has been well-documented. For example, Nenkov and Scott (2014) demonstrate how consumers exposed to whimsically cute babies or baby animals exhibited more indulgent behaviors due to self-rewarding tendencies. Conversely, encountering *kindchenschema* – a category of cuteness linked to vulnerability and caretaking – elicited more restrained consumer behaviors. Similarly, Marx (2019) notes that hikers on the Appalachian Trail often perceive encountered animals as cute, a perception that prompts attempts to touch, feed, or hold them. Harris (2000) further highlights how cute products generate impressions of enticement and vulnerability, which markets exploit to influence consumers' willingness to "adopt" these products. Ngai (2010, p. 15) traces the emergence of cuteness as a recognizable style to the industrialized 19th-century United States, where it played a role in consolidating the "middle-class home as a feminized space organized primarily around commodities and consumption".

While the consumption of cuteness emphasizes a capitalist phase where consumers explore the use value and physical presence of products, viewing cuteness as biopower necessitates examining how this aesthetic governs production. Here, surplus value is extracted from the labor of both workers and the means of production. The economy thrives on the aesthetic of cuteness, generating profits and surplus value by capitalizing on people's willingness to engage physically and emotionally with the cute in both production and consumption. Human–animal encounters, wherein animals are consistently cutified, exemplify this productive phase as human participants actively labor with and for the animals. In this context, cuteness embodies the appropriation of human labor alongside the animals' super-exploitation.

What is aestheticized through the experience of the cute – both human and animal free labor and product – appears to challenge Marx's (2000) assertion that workers experience alienation and estrangement from their products, given that they do not own the means of production. The aesthetic of cuteness fosters voluntary labor that does not rely on wages as compensation. Instead, it cultivates intimate and loving relationships between laborers and the means of production. From this perspective, the capitalist relations underpinned by cuteness may seem utopian and ideal, as cute labor appears to invert the traditional capitalist logic of appropriation and exploitation. Could the capitalist mode of production support cooperative relationships and resolve the inherent conflicts between laborers and the means of production?

In *Marx* (2000), alienation in the capitalist system manifests through four key aspects. First, workers experience alienation from the product of their labor, as the product, owned and controlled by others, becomes detached and often returns as a force outside the worker's control. Second, workers are estranged from the act of production, which becomes a compulsion rather than an expression of creativity and individuality. Third, this estrangement results in a fragmented "species-life", reducing human life to means rather than ends. Fourth, alienation fosters competition and antagonism between individuals, disrupting collaboration within human communities. Does the aesthetic of cuteness between humans and animals create experiences that resist the patterns of alienated labor? If so, how do we interpret this resistance?

To address these questions, it is essential to revisit the power structure and mechanisms of cuteness. As a form of biopower, cuteness not only reconciles seemingly paradoxical conflicts within capitalism but also reinforces and sustains existing relationships. According to Taylor (2014), biopower differs from sovereign power by focusing on the corporeal aspects of life – such as dwelling, desires, health, bodily care, and education – that traditionally lie outside the interests of sovereign authority. Biopower operates on bodies through norms rather than laws, regulates through internalized subjectivities rather than imposing authority from above via acts or threats of violence, and disperses across society rather than being concentrated in individuals or government entities.

As a form of biopower, cuteness functions and materializes through bodies. It informs social norms, shapes particular subjectivities, and disperses its influence across populations. This dynamic suggests that the alienated labor of panda fans, mediated through the giant panda, transforms when embodied, experienced as social norms, exercised across society, and internalized by individuals. Instead of producing feelings of estrangement or being overwhelmed – a hallmark of alienation – this labor can manifest as escalating pleasure. This juxtaposition of biopower politics over the body with the capitalization of the body's labor allows cuteness to emerge as a powerful aesthetic category, obscuring and legitimizing judgments against the exploitation of both human and animal labor.

In the following discussion, I return to panda fans' construction of the giant panda's cuteness as an empirical field where the aesthetics of cuteness are constructed and materialized in the form of biopower and the processes of capitalization. The goal is to demonstrate that the labor process, aestheticized through cuteness, and the associated exploitation of animals do not represent a utopian resistance against the capitalist system. Instead, they remain integral to the profit-generation and cost-saving logic of capitalism.

Embodied cuteness

Exercising biopower means that cuteness regulates and exerts control through the body. For Harris (2000), the cute evokes a sadistic desire in consumers, compelling them to control, manipulate, distort, and even deform cute commodities. Nenkov and Scott (2014) argue that whimsical cuteness primes consumers' self-rewarding instincts, driving them to "consume" what appears to be cute. Interactions involving hugging, holding, touching, and feeding between humans and cute animals (Genosko, 2005; Lv et al., 2021; Marx, 2019) are propelled by an irresistible temptation to engage physically and intimately with nonhuman animals. Panda fandom exemplifies this physicality of cuteness. In fact, precisely because the giant panda is arguably one of the cutest animals on the planet, fans' approach to the panda often involves aestheticizing the animal to the limits and boundaries of its physical form. As Xu Bin (M, 31, dedicated fan) observes:

> Sometimes, when it's asleep over there, it feels like a different kind of feeling. Or when it's moving around or doing something else, you just always feel that each time . . . no matter what it's doing, the level of its cuteness varies. Like when Huahua is standing there, I think it looks so well-behaved! It's round, with a round face and small ears. And when it walks up to the front and looks at the visitors, it also looks very well-behaved. It's just . . . just when you look at Huahua, for example, when it lifts its head, or its gaze, or sometimes when it likes to smile a bit, you find it very cute. I just feel that . . . after all, Huahua looks quite chubby.

For Xu Bin, a dedicated panda fan who visits the Panda Base daily, panda Hua Hua embodies cuteness in its entirety. Hua Hua's living form captivates Xu Bin, who closely observes the animal and concludes that everything it does – its movements, expressions, posture, and gait – is irresistibly cute.

Song Ya (F, 27, junior fan) shared her experiences of the panda's cuteness with me:

SY: I went over these to take photos of the pandas, and they were so . . . so cute. And, um . . . last week was my first time at the Panda Base. Before, I enjoyed photographing my friends and family. But later, I thought pandas were also very cute. So, I bought a special lens just to take pictures of pandas.

YG: Oh, so you specifically go there to take photos of pandas?

SY: Yes, exactly.

YG: OK. So how did you come to like Hua Hua? What was the process?

SY: Well, at the beginning, I though it looked kind of silly and adorable. And its shape is very cute as well. It's just Hua Hua looks even sillier and cuter.

YG: What's your story with the giant panda?

SY: The story of the giant panda? Well, last week . . . two weeks ago, I went to see the pandas with my friend. And after seeing them, I thought they were so cute. One even stuck its tongue out, which was super . . . super adorable. I really wanted to take one home, but also . . . I really wanted to pet it. But you can't, so that's how it is. I really wanted to pet it, really wanted to.

Song Ya's story was not unique. When I first met the 27-year-old freelancer who taught guitar lessons to children, she had just begun her journey as a serious panda fan a week earlier. She explained that her growing interest in the giant panda would soon make her a dedicated fan, committed to visiting the Panda Base whenever possible. To support this newfound dedication, she invested in a camera to capture better and more detailed images of the pandas. During my fieldwork, the presence of a camera often distinguished dedicated fans from junior fans. For Song Ya, the camera became a tool to intensify her physical connection and desire for the giant panda.

Song Ya was particularly struck by the giant panda's cuteness – she recounted being overwhelmed when she saw one stick its tongue out. For her, the cuteness of the giant panda was not only an abstract concept but also an embodiment of the animal itself. Cuteness, in this sense, was defined through the panda's movements and physical form.

Zhu Peng (M, 42), another dedicated fan, said that he had been visiting the Panda Base daily for the past 7 months to photograph pandas. He believed his encounters with Hua Hua were destined by fate:

> Look, look, there are so many people watching. Running for 3 to 4 days and they still can't see pandas, just see their butts. Isn't that right? But some people, they . . . just accidentally end up seeing other pandas, isn't that right? It's all about fate. Seeing Huahua, I personally think, is about fate.

Zhu Peng explained that he felt lucky to have seen Hua Hua's face and was immediately drawn to the bear during his visit. Despite this, he admitted that Hua Hua's younger brother, Yeye, was the most handsome bear he had ever seen. Zhu Peng also noted that his affection for pandas was universal:

> Ah . . . not really. I don't have a particularly strong affection for any specific panda. I just like to see them in general . . . ah, it's just a broad kind of liking. I feel that any of them are very cute.

For Zhu Peng, the experience of seeing Hua Hua's face rather than her butt was the trigger for his passion for the animal. This physical proximity – being able to see Hua Hua's face – drew him closer to the panda. Yet, the middle-aged man emphasized that he could experience the same level of cuteness with any panda. To him, the cuteness of the giant panda extended beyond individual pandas and represented the species as a whole.

Hou Yi (F, 42, junior fan) expressed that even the panda's feces could evoke cuteness:

> "Right. I . . . I want to clean up after it, I would even pay to clean up after it. Yes".

This statement illustrates the completeness of the panda's cuteness – even her feces were imbued with value, inspiring fans to feel a stronger connection to her under the charm of cuteness.

Ma Juan (F, 40, dedicated fan) echoed this sentiment:

> So cute! No, no. When it was born, we just thought it was too cute! Especially cute, so cute! And it just makes us feel . . . our mood is quite happy. Because it . . . sometimes it gets lively, or . . . from when it was small to now, you see that it's always so well-behaved! So adorable! Very cute, all of them are very cute, and I really like them. I think each one has its own unique characteristics.

Both junior and dedicated panda fans consistently viewed the giant panda as a "body of cuteness". Ma Juan's reflections added a temporal dimension to this corporeality by linking the panda's cuteness to its birth, demonstrating how the animal's appeal was rooted in its lifecycle.

This connection between cuteness and birth exemplifies how biopower operates at its fullest potential. Foucault (1990) argued that biopower constructs categories like sex and sexuality through bodily experiences, starting at birth, shaping contemporary subjectivities. Similarly, notions of race or ethnicity often rely on visible, physical markers that manifest at birth (Bamshad & Olson, 2003). These biological "facts" form the foundation of socially constructed categories and influence subjectivities. Foucault (1990) further demonstrated that once categories like sex are established, attempts to cross these boundaries – such as through transgender identities – invite discipline, oppression, and violence. Similarly, racial and ethnic identities, rooted in birth, continue to shape social experiences (Bhabha, 2015). Chow (2002) observed that visible and genetic signs of "ethnic difference" fix ethnic identities, tethering individuals to these constructs through both physical markers and cultural practices.

By extension, once the giant panda is cutified, her existence becomes inseparably tied to her cuteness. Panda fans have demonstrated that the panda cannot escape this construct – her gait, appearance, birth, feces, and movements are all perceived as inherently cute. Should the panda deviate from this construct and cease to be cute, she risks rejection or even existential threats, much like many non-cute, non-charismatic species already face.

Nicholls (2012) observed that the giant panda has become "virtual", her widespread popularity detaching her from her natural habitat and lesser-known wild counterparts. Songster (2018) supported this notion, citing Nathan Yaussy's classification of the giant panda among "Endangered Ugly Things", alongside species like blobfish and Ganges river dolphins. For Songster, the panda's excessive focus as an iconic species has distanced her from the broader animal world. While the exercise of cuteness and its biopower may have aided the giant panda's conservation, this process mirrors constructs like gender and ethnicity, which are grounded in corporeal facts and perpetuate social hierarchies. This framing of cuteness, though it contributes to the giant panda's survival, warrants scrutiny for the social structures it upholds and reinforces.

According to Foucault (1990, p. 125), the deployment of biopower can control and regulate populations because "it is not simply a matter of economy or ideology, it was a 'physical' matter as well". Biopower "implies a physical proximity and interplay of intense sensations" (Foucault, 1990, p. 44), requiring a constant, attentive, and curious presence for its exercise. Interactions with the giant panda serve as both an effect and an instrument of biopower. While the cuteness of the giant panda draws consistent attention and curiosity, non-cute species continue to suffer from neglect due to their lack of appeal. If cuteness has saved the giant panda from extinction, it has simultaneously imposed limits on the survival of other, non-cute species.

As Colléony et al. (2017) observed, human preferences for charismatic species often overshadow the animals' conservation needs or endangered status in biodiversity projects. This focus on species charisma also contributes to biological invasions by shaping human attitudes toward different species (Jarić et al., 2020). Moreover, the overwhelming attention directed at cutified species like the giant panda creates challenges for the animal itself. While Nicholls maintains a moderately positive perspective on the creation of the virtual panda, Songster (2018) expresses greater concern about the panda's primary identity as a wild animal. Songster argues that isolating the panda from the broader animal kingdom undermines efforts to establish its existence as a wild species and to preserve its diminishing natural habitat. This is how cuteness exercises biopower across species, paralleling the dynamics of race and gender in human society and necessitating a critical examination of the panda's embodiment of cuteness.

The cuteness of the giant panda functions as a currency circulating through the panda fandom, forming the material basis and point of exchange on which social relations and affective economies are facilitated. Ahmed (2004) defines affective economies as systems in which emotions circulate and bind communities. Similarly, Lehmann et al. (2019, p. 142) argue that affective relations and economic processes share a structural affinity, to the extent that the mechanisms of capitalism and colonialism can be understood as "a theory and practice of affect". In this light, panda fans' exploration of the bodily sensations of the giant panda and their creation of intense emotional experiences through the animal's cute imagery closely mirror capitalist relations. The cutification of the giant panda becomes an economic effort to exploit the animal as a unique means of production. This relentless exploitation pushes the bodily limits of the panda: Cameras capture the precise texture of her fur, the patterns of her patches, her movements, and even her feces, which are imbued with value. However, cuteness aestheticizes this exploitation by tying it to the panda's corporeality. If the cuteness is perceived as innate to the giant panda, it implies there is no need to justify what people take from the animal.

This logic becomes even clearer when considering parallels with concepts like gender or ethnicity. Substitute "cuteness" with gender or ethnicity, and the implications become clear: If femininity is innate to women, they are obliged to perform feminine gender roles in a patriarchal society. If a person is born Chinese, they are expected to live and act within the boundaries of that identity. Numerous studies in the fields of gender and ethnicity have exposed the falsehoods inherent in such claims. Similarly, cuteness can serve as a vantage point for rethinking our social relationships with animals like the giant panda.

Confession and transformation

The interview with Lu Qian, a 36 year-old female lawyer with a postgraduate degree and a dedicated panda fan, was easily one of the most impressive conversations I had with panda fans. Starting at the tail of the queue outside Hua Hua's enclosure, the interview lasted about 2 hours when Lu Qian found it was finally her turn to meet Hua Hua. I had to spare 3 minutes for her to see the panda, and it was one of the few times when I queued long enough to meet Hua Hua. A native of Chengdu, Lu Qian used to visit the Panda Base with her mother one or two times a year. When social media first introduced Hua Hua's posts to her, Lu Qian took a few glances but did not invest more of her time in the panda. However, during the pandemic outbreak of summer 2022, Lu Qian was instantly attracted to videos. "I felt Hua Hua was cute and my mood was gone", Lu Qian said. She purchased a camera lens for Hua Hua despite having no experience in photography before. She made the following introduction of Hua Hua to me.

YG: Hua Hua . . . Is Hua Hua different from other pandas?

LQ: In what way is she different? Her personality is different; she has a great personality. She is very . . .

YG: Do other pandas have personalities that aren't great?

LQ: Hua Hua has a good personality. No, I wouldn't say others have "bad" personalities. It's just that compared to others, Huahua's personality is more . . . maybe more calm, more content. She gives you a feeling of peace when you see her. She's not so restless.

YG: That feeling of peace is?

LQ: When you see her, she's always doing things slowly and leisurely. For example, other pandas eat very quickly. She eats more slowly, maybe because of her bite or her personality. You can see her eating for a long time. People think, "Oh, no matter what happens, Hua Hua is still there".

YG: She's always eating?

LQ: Yes, she's always there. It makes you feel like nothing major has happened, like things haven't really changed. Even if bad things happen, you feel like they will pass, and it's not a big deal because she's still there.

YG: So, how has your life changed since you started loving Hua Hua?

LQ: Hmm . . . maybe I don't feel as anxious as before. I look at her every day, and it feels like nothing major is happening. When I'm in a bad mood, seeing her makes me feel better. My work is very demanding, and the pressure is quite high.

YG: OK.

LQ: Before, I even posted on Weibo that before I knew Hua Hua, I never cared about my surroundings. For example, I never noticed the flowers and plants in my neighborhood. But after seeing Hua Hua playing with a chrysanthemum in her yard, I thought it was beautiful. Then I noticed the chrysanthemums in my own neighborhood. I realized my neighborhood is well-maintained and beautiful. I took photos and shared them on Weibo, realizing I never cared before because I was only focused on work. Before, I would go home and just sleep. Now, I make an effort to visit the base to see Hua Hua. My enthusiasm for work isn't as high as before. Life isn't just about work anymore. If it is, you often end up feeling anxious. Because work is never-ending and it comes with many challenges. You face many unpredictable situations. These unpredictable situations naturally cause anxiety. But now, I feel much better. I spend a lot of time with Hua Hua every day. Even if I just visit the base without going to her yard, I feel better. When I do see her, I feel even better. Even if I don't see her, I still feel good. Before knowing Hua Hua, I never cared about the little things around me. But after knowing Hua Hua, I find the world to be quite beautiful. At least, that's how it is for me.

Lu Qian's story with Hua Hua narrates the transformative journey of a professional woman through her encounter with the panda. What drew my attention to Lu Qian's conversation was her detailed and articulate discussion of this transformation, which highlighted how pandas boost one's mental well-being – a recurring theme in interactions between fans and the animal. Lu Qian's self-education and cultivation exemplify how the biopower of cuteness operates through social norms, employing techniques of confession and self-regulation.

Foucault (1990) contrasts biopower with sovereign power, noting that the latter exerts control and manipulation from above, primarily through laws. Biopower, on the other hand, operates through social norms that employ "continuous regulatory and corrective mechanisms" (Foucault, 1990, p. 144) using tools such as legal sanctions, medical diagnoses, pedagogical controls, and moralistic rhetoric. The aim of biopower is to ensure population growth, maintain labor capacity, and reinforce existing social relations. Lu Qian's transformation into a panda fan illustrates how a professional woman normalizes herself within these frameworks, allowing biopower to reintegrate her into the productive population and workforce.

Lu Qian's fanship with Hua Hua began when she discovered that the panda could lift her mood. This connection was not imposed through legal or formal means but, as Foucault (1977b) describes, "infiltrated" her life through her childhood visits to the Panda Base with her mother and her exposure to countless panda videos on social media. These daily interactions with the giant panda, which offered solace, ensured that the animal played a pivotal role in helping her address her depression during the pandemic and manage anxiety related to her work. When Lu Qian recognized deviations in her demanding professional life – such as pressure, low mood, or a sense of being out of control – the presence of the giant panda allowed her to medicalize these irregularities with accessible corrective measures. Seeing the panda provided her with a way to alter her emotional state, fostering peace and calmness.

As a well-educated woman, Lu Qian was adept at exercising pedagogical and moral control over herself. She quickly reflected on her experiences of depression and anxiety, aware of their potential negative impact on her productivity. Her fanship with the giant panda offered a mechanism to maintain a positive state of mind and productivity, equipping her with tools to understand and manage her emotional challenges.

After the giant panda helped restore her to a more positive and stable mental state, Lu Qian was better able to navigate her professional responsibilities. The panda neutralized and absorbed her unproductive and negative emotions, re-establishing her engagement and commitment to her career. Interestingly, Lu Qian's work life and panda fanship appeared to exist on two contrasting extremes: Her demanding, fast-paced, and unpredictable work routines stood in stark contrast to the slow, serene, and unchanging atmosphere surrounding the giant panda.

Although Lu Qian claimed that her interest in Hua Hua reduced her enthusiasm for work and signified a shift in priorities, this shift should not be viewed as a withdrawal from biopower's influence. Instead, her entry into the panda fandom underscores her deeper integration into biopower's mechanisms. Through her engagement with the fandom, Lu Qian developed an appreciation for the beauty of small things, such as plants and flowers, and strengthened her connections to her neighborhood and community. The panda fandom thus provided pathways for further social integration through the ideals of a shared community.

What stands out in Lu Qian's transformative journey is her use of confession to articulate and construct her sense of self. Foucault (1990, p. 61) describes the role of confession and its connection to subjectivity in the following terms:

> One has to be completely taken in by this internal ruse of confession in order to attribute a fundamental role to censorship, to taboos regarding speaking and thinking; one has to have an inverted image of power in order to believe that all these voices which have spoken so long in our civilization—repeating the formidable injunction to tell what one is and what one does, what one recollects and what one has forgotten, what one is thinking and what one thinks he is not thinking—are speaking to us of freedom. An immense labor to which the West has submitted generations in order to produce—while other forms of work ensured the accumulation of capital—men's subjection: their constitution as subjects in both sense of the word.

Foucault (1990) suggests that the exercise of biopower produces "subjects" who engage in confessional acts, scrutinizing and disclosing their identities and lives. Lu Qian's journey exemplifies this confessional process as she examined her old self – focused solely on work and neglecting her surroundings – and her transformed self, which embraced a work–life balance and an appreciation for the small joys in life. Through her participation in the panda fandom, Lu Qian narrated how the giant panda, particularly Hua Hua, helped her address the challenges her old self faced. The presence of the panda drove her to focus on aspects of life she had previously overlooked, enabling her to become what she described as a better version of herself. This confessional reflection enabled Lu Qian to investigate her thoughts and actions, comparing her past and present states and providing her with the tools to correct perceived abnormalities. Hua Hua's image became central to Lu Qian's construction of subjectivity, serving as a mirror that guided her understanding of who she was and what she should strive to become.

The biopower of cuteness, as Lu Qian experienced it, operated not only through external social norms but also through her voluntary self-surveillance and application of self-techniques. Her transformation was centered on her own sense of agency, highlighting her willing adoption of these practices. Lu Qian actively sought self-improvement without external enforcement. For example, while her mother frequently visited the Panda Base and social media initially recommended panda-related content, these influences did not immediately spark her interest. Instead, Lu Qian's panda fandom began when she consciously connected her emotional state and sense of self to the giant panda's image. The panda's presence became a catalyst for her self-monitoring, prompting meticulous evaluations of her body, mood, health, and overall well-being. However, the broader implications of this transformation – such as her increased productivity and renewed labor capacity, which align with the goals of biopower to manage and control populations – were absent from Lu Qian's narrative.

The interplay of social norms, self-reflection, and subjectivity construction that biopower employs through the cuteness of the giant panda becomes especially significant when considered in light of Marx's critique of capitalist relationships. According to Marx (2000), a professional like Lu Qian might typically experience alienation in her relationship with the giant panda, which, as a means of production, organizes and compels her labor. In this capitalist framework, her interaction with the panda would result in products beyond her control, created through mechanical and rigid processes, while also fostering a sense of alienation from her community. Although Lu Qian's professional life reflects these aspects of alienation, her panda fandom appears to serve as a counterbalance to her alienated work experiences.

On the surface, the product of Lu Qian's fanship with the panda was her renewed and transformed sense of self. It seemed that this subjectivity belonged entirely to Lu Qian, crafted for her benefit. Her fanship appeared spontaneous and creative, as she decided independently when and how to engage with Hua Hua. This process was meaningful and transformative for her, granting her new perspectives on coexisting with her environment, including flowers, plants, and her community. Her deepened interest in the panda strengthened rather than weakened her connection to her neighborhood, nature, and society.

However, this interpretation of Lu Qian's experience may idealize biopower, as it overlooks the mechanisms that render such transformations possible and necessary. Lu Qian's construction of subjectivity as a product of her panda fanship provided a sense of balance and counterpoint to her alienated professional work. Yet, this raises questions about whether Lu Qian truly owned her subjectivity or if it was shaped by broader societal forces.

The need for consistent self-evaluation is not surprising, given Lu Qian's role as a professional contributing significant labor and productivity. In seeking to balance her demanding career with the joy and peace offered by panda fandom, Lu Qian embraced the aesthetic and emotional power of cuteness to maintain equilibrium. When asked whether her fanship had received support from family and friends, she shared that both her mother and husband were supportive:

YG: What do they do to support you?
LQ: As long as they don't stop me, that's support.
YG: Hahaha.
LQ: They don't find it strange. You know, many people might not understand it, right? They might think it's weird, but they don't. They think it's nothing unusual. Some of my friends even envy me.
YG: How?
LQ: They feel like they don't have anything that excites them as much as this excites me.
YG: Have you tried to get them into it too?
LQ: No, no. I think it depends on fate. They like Hua Hua too, but they just think he's cute; they won't come and wait in line for so long, especially in the summer mornings when they have to go back to work.
YG: Hmm . . .
LQ: In the summer, the zoo closes early. So if you really miss Hua Hua, you have to come early, around 5 a.m. to arrive at 6. Then you line up and maybe catch a glimpse for a short while before he goes back inside, and then you go back to work. For my friends, it's too intense and tiring.
YG: How do you keep up with it?
LQ: I don't go every day. I might go twice a week, or occasionally. I don't think I'm particularly crazy about it. I don't force myself to do it every day. I just go when I feel like it.

Lu Qian believed her passion for Hua Hua set her apart from her friends, who admired her interest in the animal: "they don't have anything that excites them as much as this excites me." With her enthusiasm for the giant panda, Lu Qian managed to reorganize her work–life schedule, even though her friends sometimes viewed her routine as intense. Her carefully designed schedule, especially during a typical summer day, began at 05:00, including an hour-long commute and a brief 3-minute glimpse of Hua Hua before heading back to work. Despite the structured and seemingly demanding integration of panda visits into her busy professional life, Lu Qian did not consider herself a "crazy fan". Instead, she described her interest in meeting the giant panda in the early morning as driven by her personal feelings and individual spontaneity, underscoring the randomness and self-focused dynamics of her engagement with Hua Hua.

This emphasis on a subjectivity that transcended her career, daily routines, family obligations, and social connections created the conditions for the cultivation of Lu Qian's individuality. This individuality emerged from her interactions with a nonhuman species, a being with whom she could not fully communicate and whose existence was largely detached from her social and professional spheres. Furthermore, this individuality was closely tied to her experiences of pressure and anxiety within her demanding career. This raises a crucial question: How could panda fandom, rooted in alienated relations, become a meaningful and valid experience for Lu Qian's self-construction and subjectivity? If her interest in Hua Hua stemmed from an awareness of alienation in her labor, why did she frame her panda fandom as a personal, external, and seemingly unrelated activity to her work?

Lu Qian valued her panda visits and encounters because they provided a space and conditions to materialize her individuality and construct her sense of self. However, the value of this self depended on how her panda fandom enabled self-regulation, normalization, and continuous improvement – qualities essential to her career and professional life. Although juxtaposed with the giant panda, this constructed self seemed external to Lu Qian's social and professional relationships. Yet, this private sphere was where she recovered from her demanding and anxiety-inducing work environment. Thus, the self that emerged from her participation in panda fandom was both a consequence of and a solution to the challenges posed by the alienated capitalist labor process, both in its content and form.

What becomes particularly compelling is how individuality emerges under the influence of cuteness as biopower. The ways in which this subject reacts and describes itself form the materiality of the cute. Both Foucault and Marx agree that individuality is a significant and inevitable component of capitalist relations and biopower. While cuteness functions as a biopower to reinforce existing capitalist structures, the construction of self becomes both a consequence and a necessity. Panda fandom provides a distinct sphere where fans dedicate time and effort to transforming and constructing their individualities.

Normalization

In the rise of Hua Hua as a celebrity at the Panda Base, cuteness functions as a significant aesthetic strategy that normalizes Hua Hua's differences from other captive giant pandas. Due to her developmental delays, Hua Hua was unable to climb trees and exhibited a mild temperament. Additionally, her highly recognizable white fur and chubbiness further set her apart (Global Times, 2023). Within panda fandom, these differences are often viewed as contributing to Hua Hua's cuteness. For example, as Lu Qian's interview demonstrated, Hua Hua was regarded as a peaceful and calming bear, specifically because of her lack of competitive awareness. Zhao Ming (M, 28, dedicated fan) elaborated:

> Hua Hua! The top celebrity among all! A star! The top! Because Hua Hua's recognition . . . recognition . . . recognition is very high, you know. Like other pandas, if I were to see . . . other pandas, they all seem similar, right? I just can't remember their names. But as for Hua Hua, it has very distinctive features. At a glance, you can tell it's Hua Hua. It's very obvious compared to its younger brother; you can tell them apart at a glance.

As one of Hua Hua's dedicated fans, Zhao Ming shared his perspective on how people should treat Hua Hua with proper care and understanding.

ZM: Many . . . many people just say about Hua Hua . . . I mean, about the disabled kind, right. I mean, how to say it . . . just . . . just whatever. Just fundamentally . . . I mean, saying it's not good, that kind of thing . . . feels like . . . all pandas are cute, right? You can't . . . deliberately defame it. Saying things like . . . Hua Hua can't climb trees, like Hua Hua has . . . has that what . . . has defects like that. People shouldn't say it like that.

YG: So . . . then how should people talk about it?

ZM: Just say it can't climb trees. Right. If it can't climb trees, it can't climb trees, that's very normal. If there's something . . . some defect like that, can't do it, you can say it that way. Anyway, just . . . how to say it? I'm not very good at saying.

Hua Hua's inability to climb trees highlighted her vulnerability and evoked a sense of "heartache" among her fans. Zhao Ming's concern about criticism aimed at Hua Hua for her inability to climb trees was countered by a strong fan narrative that reframed this limitation as "normal". Like many of Hua Hua's admirers, Zhao Ming did not view her inability as a deficiency but instead celebrated her cuteness as a compensating virtue. Hua Hua's smaller size and perceived insufficiencies further aestheticized her interactions with other bears, transforming them into symbols of her unique qualities.

Undoubtedly, if Hua Hua were born in the wild, her inability to climb trees and her mild temperament would have severely hinder her chances of survival in the competitive natural world. However, within the controlled environment of captivity, these natural deficiencies were reframed as endearing qualities that enhanced her appeal. The normalization of Hua Hua's traits exemplifies the exercise of biopower through cuteness. Her physical appearance and temperament accentuated attributes associated with human infantile features, allowing Hua Hua to wield an extraordinary appeal rooted in cuteness. Unlike other captive pandas, Hua Hua's perceived differences were not considered pathological or abnormal but were elevated to the status of virtues, enabling her to become a top celebrity in panda fandom.

I argue that in the making of Hua Hua a celebrity bear, cuteness as a form of biopower operates at a heightened level of discipline, regulation, and control. Traits that might have been fatal in a natural habitat – such as her inability to climb trees – are transformed into sources of value, fame, and admiration in her interactions with humans. In the realm of biopower, life is not excluded or discarded due to its deviation from the norm. Instead, it is regulated, managed, and interwoven into human society. Hua Hua's celebrification exemplifies the triumph of biopower over the laws of natural selection. Her life is not marginalized but actively embraced and redefined through the aesthetics of cuteness. In its drive to "make life", the biopower of cuteness reshapes and reimagines Hua Hua's existence as an aesthetic project.

Foucault's (1988, p. 49) concept of the "aesthetics of existence", introduced in his later work, provides a framework for understanding this normalization process. As Foucault (1988, pp. 49–50) suggests, as power shifts from sovereign authority and strict obedience to dispersed and disciplined biopower, "the search for an aesthetic of existence" becomes necessary. He elaborates further:

> indeed, in most of these analyses, people are not told what they ought to be, what they ought to do, what they ought to believe and think. What they do rather is to bring out how up till now social mechanisms had been able to operate, how the forms of repression and constraint had acted, and then, it seems to me, people were left to make up their own minds, to choose, in the light of all this, their own existence.

For Foucault, the aesthetics of existence embodies the possibility of living one's life as an artwork, shaped and guided by the mechanisms and structures of biopower. Confession, as a self-examination of one's life, creates opportunities for individuals to question their actions, problematize their place in the world, and explore the means available for constructing and shaping their sense of self or selfhood (Huijer, 1999). Foucault suggests that the exercise of biopower is intricately linked with the aesthetics of existence. In the context of panda fandom, this theory implies that cuteness not only disciplines individuals but also provides aesthetic principles that fans use to make creative decisions about their relationships with pandas and the construction of their individualities.

Dong Mei, a freelancer and dedicated panda fan in her mid-30s, shared with me her distinctive journey of self-construction within the panda fandom.

YG: So, was a giant panda fan, what does it mean to you now?

DM: What does it mean? It gives me a spiritual support. Because I advocate for not getting married or having children. So, I find a kind of spiritual support in this.

YG: Hmm, is it based on the relationship between you and the pandas?

DM: Yes. I have a one-sided affection towards them, just a one-sided kind of feeling.

YG: OK. OK. Why did you choose pandas?

DM: They are very healing! Because of them, their way of life. Hmm, it's more like I have been searching . . . searching for a kind of feeling, a fourth kind of feeling besides familial love, friendship, and romantic love.

YG: And you found it in pandas.

DM: Yes, yes. That's exactly the feeling, that feeling. I can't describe it, but I feel . . . It's beyond those three kinds of feelings, another kind of feeling.

YG: OK. Do you have this kind of feeling towards flowers? Since you used to run a flower shop.

DM: Hmm, flowers are for survival. They are a business skill you need. But you need them to make a living. It's completely different; that's a survival skill, this is spiritual support.

YG: what changes have happened in your life after having this fourth kind of feeling?

DM: Hmm, I'm very happy! Sometimes very happy; sometimes, seeing pandas not doing well, I get very sad too. It's beyond these kinds of feelings. Another kind of feeling, that's the only way I can describe it.

YG: OK. OK. So, how do you view the time spent on pandas?

DM: I'm very willing. Especially willing. Even when I'm busy sometimes; I'm always willing to take a bit of time, an hour, to come over. I'm willing to see them. A few days ago, I delivered flowers to the west gate of the Panda Base. Then I came in for a round, and left through the south gate. So, I'm very willing to spend time on pandas. I like this feeling. I don't need pandas . . . I don't need pandas to do anything. Just them being there is enough.

Like Lu Qian, Dong Mei seamlessly integrated her interest in the giant panda into her work schedule while also finding the panda experience deeply enriching and personally meaningful. Dong Mei described her passion for the giant panda as a *fourth emotion* – an emotional connection beyond the traditional social ties of family, friendship, and romance. Although Dong Mei openly supported singlehood, her characterization of the panda fandom as a *fourth emotion* paralleled Lu Qian's experience. For Lu Qian, the fandom similarly offered a space outside her established social networks to construct her individuality and selfhood.

Dong Mei consciously crafted her sense of self and identity through the aesthetic of cuteness, creatively embracing the *fourth emotion* as a means to reconcile her commitment to singlehood – a position that subtly challenges prevailing social norms and traditions in contemporary China. However, Dong Mei's stance was not a straightforward rejection of traditional values such as marriage and family ties. Instead, she cultivated her relationship with the giant panda as a substitute for these relationships, demonstrating that, even in singlehood, she could experience joy and self-sufficiency through panda fandom. By aestheticizing what might traditionally be seen as a "lack" in her life, Dong Mei reframed her experience with the giant panda as a source of fulfillment and emotional richness.

After sharing two additional stories between Hua Hua and her fans, Tang Xue (F, 31, dedicated fan) highlighted how family ties added a meaningful layer to her connection with the giant panda. Tang Xue described Hua Hua's life journey as embodying virtues such as patience, self-paced growth, and moderation – qualities that inspired her in the face of China's fast-paced, high-pressure, efficiency-driven society.

Tang Xue believed Hua Hua's story offered empowerment to her and her friends. She recounted moving anecdotes, including that of a woman recovering from breast cancer and a failed marriage who found solace in watching Hua Hua daily and a rebellious young girl saved from a suicide attempt by a firefighter who shared a Hua Hua video with her. Tang Xue also shared her own struggles at the start of 2023, a period during which she lost interest in the Internet. Seeing Hua Hua on social media again after 3 months, Tang Xue was struck by the panda's growth and felt she had "missed a lot". From that moment, Tang Xue invested in a deeper, familial connection with Hua Hua:

> Yes. Actually, I've been with them since they were two years old, treating them just like children. Because its actions, behaviors, and the way it looks are very much like a little girl's; I treat it like my own kid. I feel like, yeah . . . I visit it every day. For example, with this one . . . Hua Hua. For instance, if someone feeds or something like that, because I don't want it to get sick from not eating, I care for it like my own child. And then I might contact security to see if grandpa is around and solve these problems. You feel like . . . because of your actions, you've prevented it from getting sick or something. It's this kind of emotion, feeling like you're caring for your own child.

Tang Xue added the following comments to her shared kinship with the panda:

> It's not really for any purpose. It's just . . . also out of interest and hobby. It's like, you feel like, oh, you're growing yourself, and it's growing too. Most of the time, you don't think of yourself as its fan. It feels like, um . . . growing up together, living together kind of feeling. And then you experience life together. Yes, exactly. You feel like, you see what it goes through every day, and you adjust to those things you encounter in life, it's like mutually healing.

Her aspiration for personal growth in her panda fandom was Tang Xue's determination to "making life". After she became aware of her depression and personal lows at the beginning of 2023, Tang Xue actively sought interaction with Hua Hua to allow her to transform her own life. She also collected some of the life-transforming stories based on Hua Hua and was determined that she, too, could "grow up" with the giant panda. Here, we can see that not only Hua Hua's differences with other bears but also crisis taking places in the human world, as extreme as suicide, divorce, self-abusing, cancer, and adolescent rebellious phase, can be sufficiently treated, pacified, and mended through the rhetoric of the cute. This influence and possibility of cuteness stems from its biopower, one of the most bounding power forces in the contemporary society.

Whether as an extra-work activity, a *fourth emotion,* or a familial kinship, it is notable that personal growth and the construction of self were considered meaningful and fundamental purposes behind fans' interest in the giant panda. This creation of selfhood consistently played a healing and restorative role, filling gaps fans often found lacking in traditional social relationships. Indeed, Tang Xue's examples suggest that some existing social relationships – such as family, kinship, or marriage – can be so distressing that fans sought solace and renewal in the panda fandom. Through this space, fans untangled and mitigated the impacts of such relationships. The emphasis on a self that could grow, transform, and validate itself through confessional self-inspection provided a path for fans to reconcile with existing social and capitalist structures, even as the panda's cuteness appeared to offer a departure from those aims of personal improvement and transformation. Cuteness normalized what had become aberrant or deviant in fans' lives, providing a space where they could aesthetically reframe their existence.

This chapter builds on the conclusions drawn in Chapter 3, further exploring how cuteness exercises its biopower in panda fandom. It examines how capitalist relations and biopower intertwine within the fandom, revealing how fans' conflicts with the giant panda, as a means of production, are reconciled through biopower's techniques. Cuteness exerts its influence as a powerful form of biopower, contributing to what I call the *cute existence*. This chapter demonstrates how cuteness materializes through the corporeality of pandas, circulates capitalist relations through fans' confessions and constructions of subjectivity, and deploys norms to aestheticize and regulate life.

The panda fandom illustrates biopower's strategies for sustaining and maintaining the economy and its cycles. Rooted in biopower, cuteness has emerged as one of the most dominant aesthetic categories in contemporary society, shaping and aestheticizing the experiences within the panda fandom. The fandom comprises a community of fans who connect to the giant panda through biopower's structures and mechanisms. However, because of its *bio*-origins, the aesthetics of cuteness are inherently anthropocentric, rooted in the image of human infants. This approach to biopower's task of "making life" suggests that, despite their embodiment of cuteness, giant pandas remain a super-exploited means of production in the panda fandom, where they ostensibly occupy a central place.

In this anthropocentric framework, fans' connections with pandas are conditioned and limited, even as fans profess selfless and devoted passion toward the animals. The panda fandom is not a space where pandas are fully elevated or integrated into human society. Instead, panda love has facilitated an economy that perpetuates the giant panda's super-exploited status, contradicting any initial idealization that the fandom represents a harmonious space where Chinese fans enjoy life alongside this beloved animal.

In the next chapter, I investigate fans' experiences and interactions with pandas in search of hope for a more-than-human, multispecies world that fans have envisioned or imagined. This pursuit has been a guiding inner voice throughout my conversations and discussions with panda fans in the field. If Foucault's observation that power and resistance always coexist is accurate, the panda fandom is also a space where the animal's super-exploitation can be critically reexamined and reimagined.

9

BIOPOWERED

"I feel like they love us"

One question I consistently posed to panda fans was about what attracted them to the animal. I was, to some extent, concerned about my own lack of devoted interest in the giant panda. I felt that if one of the fans could convince me of their attachment and love for the animal, I might be able to experience the same. Moreover, this formula of panda love could serve a greater purpose – if we all learned to love animals in this way, a biodiverse Anthropocene might be within reach. The panda fandom might hold the key to solving the world's problems if the love for pandas could be extracted, spread, and taught. In my fieldwork, I encountered two types of narratives between the panda fans and their beloved animals. The first narrative affirmed the animal's independence from the fans, with panda love portrayed as an unrequited devotion from the fans. Grounded in reason and logic, this narrative rejected anthropomorphizing the giant panda as a thinking or feeling being. In contrast, the second narrative emphasized reciprocity between the fans and the giant panda. Fans holding this view firmly believed in the animal's ability to interact and respond to them, often sharing these beliefs with great enthusiasm.

My research showed that fans' interactions with the pandas are primarily visual experiences. Since 2018, the "hold-a-panda" program was permanently suspended in Sichuan by the Forestry Department to reduce zoonotic transmissions as a response to the pandas suffering from intimate contact. From that point on, fans' contact with the pandas has been restricted to viewing the animals from a distance at the enclosures. This means that fans' interest in the animal could be both unrequited and no-contact, despite studies suggesting that people often desire intimate contact with cute objects (Dale, 2017; Genosko, 2005; Harris, 2000). As an embodiment of cuteness, fans have always longed to hold, feed, or touch a giant panda. What becomes interesting, however, is that fans' desire for intimacy with

DOI: 10.4324/9781003511229-10

the animal, despite the curtailment imposed by the distance of the panda enclosures, does not diminish. Instead, the distance seems to suspend fans from treating pandas as mere pets, and, as the saying goes, "absence makes the heart grow fonder" (Bhattacharya et al., 2017; Hamwey et al., 2019).

As a result, fans' attachment and devotion to the giant panda must be understood as a relationship conditioned by specific cultural, social, economic, and environmental factors. This specific historicity allows the panda fandom to be experienced as passion, feeling, and emotion in the living bodies of a global population interacting with the animal today. However, this liveliness does not necessarily justify the animal–human entanglement. Instead, the panda love warrants close scrutiny and a deeper understanding, with the hope that this examination can pave the way for a more-than-human world to emerge.

Always more-than-human

In his seminal study on a more-than-human world, Abram (2012, p. 156) writes, challenging the Western philosophical tradition, about a sensuous reciprocity between the human and the natural world in these words:

> The human mind is not some otherworldly essence that comes to house itself inside our physiology. Rather, it is instilled and provoked by the sensorial field itself, induced by the tensions and participations between the human body and the animate earth. The invisible shapes of smells, rhythms of cricketsong, and the movement of shadows all, in a sense, provide the subtle body of our thoughts. Our own reflections, we might say, are a part of the play of light and its reflections. "The inner—what is it, if not intensified sky?"
>
> By acknowledging such links between the inner, psychological world and the perceptual terrain that surrounds us, we begin to turn inside-out, loosening the psyche from its confinement within a strictly human sphere, freeing sentience to return to the sensible world that contains us. Intelligence is no longer ours alone but is a property of the earth; we are in it, of it, immersed in its depths. And indeed each terrain, each ecology, seems to have its own particular intelligence, its unique vernacular of soil and leaf and sky.

Based on Merleau-Ponty's phenomenology, Abram's idea suggests that the human mind is inherently embodied and situated within the natural world. Human sensory experiences are deeply intertwined with the environment around them, making the human mind and body part of a sensible world. In this reciprocity, the human mind is not an isolated, ethereal entity within the body but emerges from, and is shaped by, sensory interactions with the environment. The sensorial world provokes and maintains the activity of both the individual and the mind. This connection between the internal psychological world and the external perceptual world helps blur boundaries, turning the mind inside-out. Abram's more-than-human worldview implies that a holistic understanding of self ties well-being, health, and the construction of self to individuals' engagement with the environment.

While Abram's notion seeks to reconnect environmental consciousness and human internality to nature, challenging the mind–body dualism in Western philosophy, his recognition of the sensuous human being resonates with Foucault's emphasis on the constructed self, shaped by the discipline and influence of biopower. Although Foucault (1990) focuses on the power structures in human society, Abram approaches the human individual through the "power" of nature and the sensuous world. Thus, for both philosophers, the self is constructed and influenced by external environments, whether societal or ecological.

In earlier chapters, I discussed how Foucault demonstrates that the construction of individuality and selfhood has always been "more-than-human" in the sense that biopower disciplines and regulates the individual and population. Abram suggests that perhaps biopower has always been the mode of interaction and connection between human individuals and the ecological world. Interestingly, while Foucault cautions us about biopower's influence, control, and manipulation in the construction and domination of the human self, Abram advocates for a different stance – one where the individual embraces the biopower of nature. The body's senses, experiences, and feelings need to synchronize with the more-than-human world.

How the two more-than-human worlds – the ecological and societal – hold contrasting influences and techniques, yet both mobilize human individuals through their biopowers, is a question deserving further examination in a future project. In this chapter, I take a step toward connecting these two "more-than-human worlds". While Foucault's concern lies in how the human self searches for emancipation, liberation, and freedom from the manipulation of biopower in the societal sphere, Abram asks how human individuals can be further "biopowered" by the nature's more-than-human world. Whereas Foucault focuses on resistance against biopower, Abram proposes a surrendering vulnerability to being influenced, embodied, and moved by the external world. How can the human self simultaneously resist and surrender to biopower? If Abram's more-than-human world calls for the individual's vulnerability to biopower, how can this sensuous human individual echo Foucault's study of disciplined human bodies?

In the following sections, I explore how panda fans articulate their interactions with pandas to show how both the ecological and societal biopowers structure and manipulate human individuals in similar ways.

Being there

Gao Lei (M, 28, dedicated fan) explained his attachment with the giant panda to me:

YG: So, do you have any emotional interaction with the panda when you film her every day?

GL: I can give one-way. I don't need to know the panda's inner thoughts, nor do I need a lot of interaction with the panda.

YG: OK. So, do you need to know what the panda is thinking?
GL: No, I don't.
YG: Do you think the panda is happy every day?
GL: As long as I believe she's happy every day, that's enough.
YG: Your belief. Is that enough?
GL: Because whether the panda is happy or not, I actually don't know.

Gao Lei traveled from Guangdong Province to Chengdu (1,273 km) each month to photograph pandas at the Panda Base. He shared with other 29 panda fans the idea that their relationship with the giant panda was "one-way" in the light that the giant panda's significance is their shared existence with the fans, or, in the words of a number of philosophers, the panda's "presence" (Heater, 1992; Riva et al., 2003), "being there" (Brown & Ryan, 2003; Clark, 1998; Ratcliffe, 2005), "being in the world" (Dreyfus, 1990; Mulhall, 2014), and "in the moment" (Ruedy & Schweitzer, 2010).

With Dong Mei (F, 36, dedicated fan), I further explored the fans' appreciation of and dedication to the panda's presence, focusing on the idea of the panda's "just being there".

DM: I like this feeling. I don't need it . . . I don't need the panda to do anything. It just . . . just being there is enough. I don't need it to respond to me. I'm content with giving unconditionally by myself, that's enough.
YG: You're satisfied? You really enjoy it?
DM: Yes, exactly. It . . . it gives you satisfaction, a kind of spiritual comfort, that's it.
YG: OK. So, in about . . . about 5 years, what do you think your relationship with the panda will be like?
DM: Relationship? It's still this kind of feeling! I'll always love them, and I'll love them more and more deeply.
YG: OK. Actually, you don't expect them to do anything in return, it's just you giving unconditionally?
DM: No, I don't need that.
YG: Just you loving them, that's enough?
DM: Yes, as long as they don't get sick. As long as they don't die, I'll always . . . always . . . love them.
YG: Haha, you're really persistent!
DM: It's not that bad, I believe most people are . . . would be like this just because you like them. I also . . . I know some fans who have been fans for ten . . . over 10 years. They really like pandas. Every year, for holidays, they fly over here just to see the pandas. I also know a lady. She especially loves a particular panda; she rents a place near the base. She just stays here to watch them. Sometimes for 2 or 3 months; then she goes back to take care of things, and then comes back again.
YG: Hmm, that's quite deep, very deep connection you have. Hmm . . . so do you think, for example . . . Hua Hua, does she need these fans?

DM: She doesn't need fans. It's the fans who need her. No. For Hua Hua, she's just a simple panda. She eats, sleeps, drinks every day, that's enough. She doesn't need these fans, it's the fans who need her. It's the fans who can't do without her. Who can't she do without? She can't do without Grandpa Tan (note: Grandpa Tan is Hua Hua's keeper). Yes. This is actually just a one-sided thing for the fans.

YG: OK. So, these social attributes and things; for Hua Hua, are they a burden?

DM: A burden? Not really. Because . . . as long as Hua Hua is kept in captivity, living here, every day she is just eating and drinking. Do fans coming to see Hua Hua make her happy? Does it make her sad? No. It's just that, for humans, people need Hua Hua more. It's not . . . it's not Hua Hua that needs humans more.

YG: OK. So Hua Hua doesn't think about these things?

DM: No, she doesn't. She doesn't need these human social attributes. These are all things humans assign to her. They're not things she wants.

YG: Hmm. Do you think, as a fan, as someone who loves them; do you think it's possible for humans and pandas to communicate?

DM: Can humans and pandas communicate?

YG: Yes. The relationship between you and Hua Hua, you and . . . the panda Yuan Run? Is it just one-sided?

DM: Yes, exactly. I don't need any response from them. It's just, they're there. They . . . they're there. They are my spiritual support; just seeing them there, I'm very happy.

YG: OK. I see what you mean.

DM: I don't need them to do anything, nothing special.

While delving deeper into the conversation, my "interrogation" did not lead Dong Mei away from her claims that the panda's role was simply "being there" and that she did not expect the pandas to provide or do anything in return. I was, clearly, trying to understand what initially hooked Dong Mei to the giant panda. In my opinion, it was very likely that the giant panda had done something special to attract and transform Dong Mei, as well as other fans. What was the secret between the pandas and the fans that I did not yet understand?

In one of his fieldwork days in Indonesia, a sudden thunderstorm drove Abram (2012) to take refuge in a cave. While waiting in the cave, his attention was drawn to a spider weaving webs at the cave's mouth. This close observation of the spider, as Abram (2012, p. 12) reflected, had since filled his encounter with the animal with a "feeling of great strangeness and awe". Abram (2012, p. 12) further described the spider's presence in these words:

> To be sure, insects and spiders are not the only powers, or even central presences, in the Indonesian universe. But they were my introduction to the spirits, to the magic afoot in the land. It was from them that I first learned of the

> intelligence that lurks in nonhuman nature, the ability that an alien form of sentience has to echo one's own, to instill a reverberation in oneself that temporarily shatters habitual ways of seeing and feeling, leaving one open to a world all alive, awake, and aware. It was from such small beings that my senses first learned of the countless worlds within worlds that spin in the depths of this world that we commonly inhabit, and from them that I learned that my body could, with practice, enter sensorially into these dimensions. The precise and minuscule craft of the spiders had so honed and focused my awareness that the very webwork of the universe, of which my own flesh was a part, seemed to be being spun by their arcane art.

Abram's experience with the spider and its presence is complex and dynamic as the philosopher attuned himself to the spider's "being there". The spider introduced Abram to a multitude of dimensions that he had not previously known. Abram, focusing on his senses, formed his understanding of nonhuman intelligence and sentience, prompting a deeper, spiritual connection with nature. The encounter with the spider moved him beyond his own mind and made him aware of the nonhuman creature's own form of intelligence. This awareness disrupted his established perceptions and emotions. His recognition of the presence of the nonhuman other was highlighted when he encountered trees (Abram, 2012). In this connection with trees, Abram further notes that the tree is not limited to what he saw personally but also includes a shared perception by other beings, such as birds chirping and nesting in the tree. The tree existed before his observation and would continue to exist after he looked away. The presence of the spider and the tree showed Abram that there are dimensions and intelligences that he can connect to through his senses, and these can exist beyond and beside his connection to them.

I want to bring Abram's encounter with the spider and trees into a conversation with Dong Mei's attachment to the giant panda. This means that Abram's articulated and complex reflection parallels Dong Mei's enjoyment and satisfaction in the panda's "just being there". It also explains fans' need for the giant panda, rather than the other way around. Dong Mei's need for Hua Hua lies in the fact that the panda lives her own animal life without acknowledging the interference of her human fans. Like Abram's encounter with the spider, Dong Mei's love for Hua Hua links her to the spirits and nonhuman intelligence residing in nature. Hua Hua's existence speaks for the "countless worlds within worlds", but it also allows Dong Mei to connect with those worlds through shared experience and space with Hua Hua. Nevertheless, Hua Hua's world continues to evolve and transform with or without Dong Mei's participation. Hua Hua's being there, as Abram's tree and spider show, refers to both Dong Mei's inclusion and exclusion. Dong Mei can immerse herself in the reciprocity shared between herself and the giant panda, yet she also knows that Hua Hua has worlds independent of her. Hua Hua, the spider, and the tree – these presences, "just being there", for both Abram and Dong Mei – define their human experience and existence in the multiple worlds to which they belong.

In his exploration of the relationship between art and ecological consciousness, Morton (2021) stated that "all art is ecological". Morton suggests that the experience of art offers a model for the coexistence between humans and nonhumans in more-than-human worlds. As the subject experiences beauty through art, Morton (2021) writes, the subject coexists with at least one thing that is the other through noncoercive engagement. Morton (2021) explains that a person's appreciation of an artwork evokes a sense of connection with the external other, which resonates internally. In this feeling of coexistence with the other, the individual becomes aware of feelings inside themselves that were not self-generated but seemed to originate from the external other. However, the subject experiencing these feelings cannot clearly separate this feeling from themselves, nor from the object. Hence, this coexisting feeling with the other remains enigmatic, elusive, relational, and ambiguous. Morton explicates that the more-than-human world dwells in the individual's artful connection with the other. Morton's understanding offers an answer to Dong Mei's interest in the panda – there was no secret in Dong Mei's connection with the panda, as the dedicated fan was appreciating her coexistence with the animal. This appreciation of the giant panda was not about what the panda might be used for or what the animal does but, in the words of Morton (2021), an "indifference" that points to a way to "care for humans and nonhumans in a less violent way—simply allowing them to exist, like pieces of paper in your hand, like a story you might appreciate—or not—for no reason".

This "letting be", indifference, careless, and playful mode of care, according to Morton (2021), offers an ecological alternative to the survivalist mentality that focuses on immediate threats and self-protection, which often drives "too caring" conservation efforts. Too often, ecological catastrophes arise from the survival principle, in which mere existence outweighs any quality of existence. Morton (2021) argues that this survival and bottom-line mode is also the cause of the consistent exploitation of animals as a means of production in capitalist relations: The industry exists for the sake of existing because profits must be generated in spite of the suffering of the working class and other nonhuman beings. The consequence, Morton (2021) points out, is that an animal's emotions, such as hunger, are more "legitimate" and "useful" than their feelings of happiness. Morton notes that this construction of feelings – where sheer survival, like hunger, is more real than other kinds of emotional states, such as happiness – is anthropocentric. For example, the emotions of an elephant, such as hunger, pain, or fear, are often observable and can be addressed directly through caretakers. These emotions have direct, practical implications for the animal's well-being and survival. In contrast, a happy elephant does not require immediate intervention or have clear utility from a survival standpoint, as happiness in animals can be subjectively experienced and difficult to measure or address in practical terms.

Dong Mei's fanship of Hua Hua is conceived precisely through the animal's "uselessness". It is not the animal's hunger, pain, or suffering that builds up her connection to the animal. Rather, it is the animal's peaceful and relaxed happiness that grounds Dong Mei's devotion and commitment to the animal. Following Morton's (2021) observation, it can be suggested that the giant panda's survival and the threat of extinction have been the primary concerns of giant panda conservation, rather than the animal's happiness and contentment. Dong Mei's fanship resonates with the giant panda's happiness and "being there" – a less "caring" and playful feeling in contrast to the struggle for the species' survival and breeding in panda organizations around the globe (Barua, 2020; Nicholls, 2012). However, a closer look at Dong Mei's panda fanship reveals that her seeming "indifference" to the animal's survival affords a different scenario for the animal: The panda can be wherever she wants to be, in captivity or in the wild, as long as she lives her panda life to her full potential. Just by living her panda life, fans such as Dong Mei will care and love the animal forever. Dong Mei simply allows her panda to be a panda and rejoices in the panda for being a panda. Dong Mei's love implies permission for the panda to move beyond her current role as a super-exploited means of production in the panda attraction industry. This possibility to imagine a different – and perhaps more "pandaly" – future for the animal, I believe, is how the fans' panda love can make a difference in the established order of ongoing giant panda conservation.

Perceiving reciprocity

A few fans shared their experiences of interacting with the giant panda during our conversations. More than once, my interviewees spoke of a "spiritual connection" they felt with the animal. For instance, Zhu Xi, a 55-year-old retired school teacher and animal activist who had become one of the most-followed social media influencers on giant pandas, described her mutuality with Hua Hua with great enthusiasm:

YG: For the giant pandas, having a group of fans like you, how do the pandas feel about it? How do they perceive it?

ZX: How do the pandas feel? The giant pandas? I think we should not only watch them. We should also encourage everyone to observe them in a civilized manner, to protect them. I think the giant panda see us as friends, right?

YG: Friends with all humans?

ZX: I don't know about other people. I visit them often. I think they can recognize me.

YG: Do the pandas . . . do they need you?

ZX: They need us! Of course they need us! You can tell they need us! They do! I think the giant pandas, actually, sometimes when they see us watching them, they seem pretty happy! For example . . . like when Hua Hua is sitting there eating and we go to see her, I think Hua Hua . . . wow! She even makes eye contact with us. She . . . yeah.

YG: She interacts with you.

ZX: Sometimes when you quietly whisper, "Hua Hua, I'm here to see you", she seems . . . I feel like she somehow knows, like there's a telepathic connection. When she looks at you, your heart just melts! Oh my gosh, it's like I'm totally melted! I feel like there's interaction, we interact with the pandas.

YG: Can they understand humans?

ZX: They can sense us.

YG: OK. So which panda do you feel the strongest connection with?

ZX: Which panda? I watch Hua Hua the most right now. I watch Hua Hua the most because you have to queue up to see her. So I watch her a lot. Hua Hua is very . . . very spiritual.

Zhu Xi's interest in the giant panda grew significantly since 2018, as the animals became deeply integrated into her life and work. By the time of our interview in 2023, she was visiting the Panda Base every day and believed that conservation education through social media was both meaningful and her responsibility. Her aim was to better understand animals and foster multispecies harmony. Zhu Xi's prolonged and established engagement with the giant panda, it seems, was reinforced by her conviction that the giant panda had reciprocal emotions and could share the interests of its fans. Although studies (de Vere & Kuczaj, 2016; de Waal & Andrews, 2022; Mendl et al., 2022) have recognized animals as sentient and emotional beings, Zhu Xi's emotional connection to the giant pandas, I believe, should not be seen as a simple justification for the panda's emotions merely because of her strong and extended bond with the species.

However, this does not mean that Zhu Xi's experience with the giant panda should be reduced to a personal or subjective description, nor should it be dismissed as a fantasy or illusion irrelevant to ongoing scientific research. It also does not imply that Zhu Xi spent 5 years hallucinating a journey with the giant panda or that her love for the animal is fundamentally illegitimate because there is no scientifically grounded "truth" in her story. Rather, this "reciprocal encounter", in the words of Abram (2012, p. 43), between Zhu Xi and the giant panda should be understood within the context of Zhu's specific human species positionality, shaped by the limits of human knowledge and language. The objectivity in Zhu Xi's panda love, in this light, lies within an adequate account of the inherently human perspective on the panda (Cantor, 2020). As Abram (2012, p. 43) notes, "only by affirming the animateness of perceived things do we allow our words to emerge directly from the depths of our ongoing reciprocity with the world". Abram explains that a reciprocal encounter builds upon the acknowledgment of the vitality of the things we perceive, and how language comes to structure and capture this profound exchange.

For Zhu Xi, the giant panda's animateness, her interactions with people, and her demonstration of interest toward tourists affirmed the panda's otherness and engaged her bodily sensations – the giant panda was very spiritual. Moreover, this shared spontaneity with the giant panda, bordering the limits of human language, can only be captured in this way. Zhu Xi had no other linguistic means to fully convey her message during our conversation. The objectivity underlying Zhu Xi's spiritual connection with the giant panda, therefore, is not found in the animal or the fan, but within me, another human being who recognized and understood this attachment in the presence of the panda fandom. Although I did not necessarily share Zhu Xi's intimacy with the giant panda, I could fully understand her feelings and emotions because there are other objects and others I reciprocated with in different moments and places through my own human positionality.

Guo Na (F, 22, junior fan) introduced a different version of her panda love to me:

YG: What do pandas experience about having fans like you?

GN: Ah, well, Hua Hua might think, "You guys are so annoying! All I want to do every day is sleep peacefully, but here you all are with your phones and cameras, snapping pictures of me. It's so annoying!" It's like being forced to work. Every day, there's a group of people shouting at her, "Hua Hua, look over here! Hua Hua, come over!" Maybe she just wants to eat, sleep, and bask in the sun, that kind of life.

YG: So you think these fans are disturbing their lives?

GN: But not entirely. Maybe after having fans and gaining popularity, Grandpa Tan will give her more delicious treats.

YG: Hahaha! So, does she need fans then?

GN: Hmm . . . It's not that the giant pandas need fans, it's that we need pandas like these to heal us. Right?

Guo Na constructed a different personal narrative for Hua Hua. To the fresh college graduate, Hua Hua had inner voices that questioned the presence of visitors – noisy, disturbing, sometimes chaotic, and rude human groups. The giant panda also benefited from her celebrity, as her appetite was satisfied when keepers offered her additional treats. Zhu Xi and Guo Na provided contrasting subjective views on the giant panda, but both shared a mutual understanding grounded in their human positionality. In fact, Guo Na was quite explicit about her observational position regarding the giant panda, namely, as a visitor who photographed and shouted at the animal. Guo Na clarified the human viewers' position and the resulting disturbance, emphasizing that it was specifically the human presence that disrupted the panda's peace.

This identification of the tourists' role draws attention to the gap between the panda and the visitors, suggesting that, as human beings, people were limited in their ability to perceive the giant panda only through the lens of tourists. Recognizing this difference, Guo Na dramatized the gap between the two species. However, in sharing this interpretation with me, Guo Na also acknowledged that this gap was part of the objective truth between pandas and their viewers. It was, after all, this limited human positionality that allowed Guo Na to make her joke – precisely because humans were confined to their role as visitors to the giant panda, they were unable to share the panda's enjoyment of her bamboo treats in the moment. Visitors were limited to observing, photographing, and attempting to communicate with the panda through human language, while the giant panda simply lived her own life.

Feng Jing (F, 30, dedicated fan) explained that she believed her time with the giant panda was worthwhile because the animals loved her back:

> I feel like they love us. I feel surrounded by them, just full of love. I feel that animals really give you a lot; even with just a look, even if they haven't really seen you. Subjectively, I feel like they're looking at me. Wow! They give me a kind of infinite halo, a very happy feeling.

When asked how she envisioned her loving relationship with the panda in the future, Feng Jing further explained:

> In the past, we constantly sought happiness from pandas, or resources, or a sense of joy. Now, we hope to contribute. They're guiding us, and we need to respond. How do we love them? For example, it's simple. It's no longer just about watching pandas; it feels like a responsibility. Hmm . . . it's like they love us, and now we need to love them back. Hahaha. That kind of feeling. It's the feeling they give off, it's our subjective perception that makes us feel they're loving us, like loving humans. They're very gentle, never hostile or aggressive towards humans, right? I think so.

Fans also describe this reciprocity as the panda's liveliness and rich emotions (Liang Mei, F, 30, junior fan), trust in humans (Wang Qiang, M, 25, dedicated fan), interests in fans (Wu Jun, M, 23, junior fan; Song Ya, F, 27, junior fan; Gao Lei, M, 28, dedicated fan; Song Yang, M, 42, junior fan; Chen Jia, F, 28, junior fan; Gao Yu, F, 69, junior fan; and Liu Yan, 19, F, junior fan), pride in becoming a celebrity (Liu Wei, M, 65, junior fan and He Dong, M, 27, junior fan), acknowledgment of individual fans (Yang Tao, M, 29, dedicated fan; Zhao Ming, M, 28, dedicated fan; and Tang Xue, F, 31, dedicated fan), love for being lively and active (Deng Ping, F, 40, junior fan), spiritual connection (Han Xia, F, 43, dedicated fan), and responses to the crowds (Li Jun, M, 18, junior fan; Zhu Xi, F, 55, dedicated fan; Ren Hui, F, 28, dedicated fan; and Meng Xue, F, 25, junior fan).

A closer look at these reciprocal relationships between the giant panda and fans reveals that pandas, as narrated and contextualized in this reciprocity, are largely circumscribed by our civilization and its technologies. From the visitors' perspective, pandas can experience human emotions through their celebrity status and interact with the crowd as they become tourist attractions. Surrounded by tourists, keepers, and other humans, fans believe the giant panda has adapted to, and fostered trust in, human participants. Feng Jing considered the possibility of giving back to the panda, believing she had received a lot from the animal. She felt this need to reciprocate was a responsibility, although how she might proceed with this remained ambiguous and elusive in our discussion. Like Zhu Xi, Feng Jing believed she had received much from the giant panda: The animal loved the fans back, looked at them, and sometimes even acknowledged their existence. When Zhu Xi reciprocated with the giant panda, she whispered to the animal as if the panda were aware of her presence. Feng Jing stated that while she looked, the pandas also turned their gaze toward her.

Abram (2012), citing Merleau-Ponty, suggests that in the exchange between humans and nonhuman others, this reciprocity of sensuous experience can be termed "participation". The example Merleau-Ponty offered was that when a hand touches a table, the table also approaches the hand because the hand is a touchable thing. The table touches the hand back and affirms that the hand is a part of the tactile world the hand exists in. Abram (2012, p. 49) notes that human beings can perceive things since they are also a part of the sensible world, and hence "the world is perceiving itself through us". Abram and Merleau-Ponty's ideas suggest that the perception of the other returns a "perceptual reciprocity", as one's surroundings are consistently observing, sensing, and watching the human participant. Engaged in this "watchful land" (Abram, 2012, p. 50), the human subject must become mindful and respectful of their surroundings.

The experiences of Zhu Xi and Feng Jing show that both fans felt touched and seen by the giant panda, with the animal affirming that the human fans were also sensible beings a panda could touch and see. This acknowledgment of one's sensuous existence, the transitive perception, and the reversible embodiments was critical in transforming how the two fans perceived their relationship with the giant panda. Notably, rather than shouting at Hua Hua, despite being surrounded by visitors, Zhu Xi opted to whisper to the giant panda – the fan was not, due to her established long-term relationship with the panda, fighting for attention from her beloved animal. Instead, her attempt to interact with the animal was, at its best, most subtle compared to tourists calling for the panda's attention. Feng Jing, feeling the panda's watchful gaze, experienced the animal's love and wanted to give back. Abram (2012) suggests that perceptual reciprocity can significantly influence human subjects and their actions as they become conscious beings dwelling in a sentient cosmos.

This sensorial affinity between the fans and the surrounding earth, through the presence of the giant panda, was also central to Yang Tao's bonding journey with a number of pandas. A Chengdu local, Yang Tao, a 29-year-old man and a dedicated panda fan for the past 7 years, worked as an editor in the newspaper industry. He shared with me how his love for the animal had deepened over time, and he referred to himself as a "guardian of pandas".

YG: How could your love for pandas grow deeper?

YT: It's just that they're so wonderful, and I can't help but feel that way. It feels like they're a kind of spiritual support. When you look at them, it's very healing. They are so pure and innocent. They're very happy, and that makes you feel the same way.

YG: How did you fall in love with Duoduo?

YT: From the moment he was born, I started visiting him. After his mother Qifu passed away, I felt even more connected to him. I wanted to protect him. When I see him, I'm happy, and I think he's happy to see me too. (Laughs)

YG: Does Duoduo recognize you?

YT: I think he does. When he comes out to the front area . . . Yes, when he comes out. He used to stay inside, but sometimes he comes out.

YG: Does he come out because he sees you?

YT: Yes, he might look at me a bit longer. That's what I feel. Since I visit him often, he might recognize me. I used to see him when he was in the Moonlight Nursery.

Yang Tao developed a bond with several pandas over time, experiencing a reciprocal intimacy with the animals that evoked feelings of purity, innocence, and happiness for him. He shared with me that he admired the pandas' optimistic attitude and relaxed lifestyle, noting that they always seemed content as long as they had enough food.

YG: How do you know they're always happy?

YT: If they're unhappy, they show it. We can tell from their expressions. They show sadness and distress through their facial expressions.

YG: From their facial expressions?

YT: Yes, exactly. I have a video of a sad panda.

YG: A sad panda?

YT: (Taking out his camera and showed) Yes, this panda looks a bit unhappy because its sister didn't come out. You can see he looks upset, right? He's clearly not happy. Normally, he's more active and happy.

YG: So their expressions show their emotions?

YT: Yes, exactly. Look at Yasong. Normally, she's lively and happy. She's relaxed and carefree. Yes. Look, she's smiling. You see? She's smiling, right? Her mouth curved up.

YG: Yes.

Yang Tao showed me photos of giant pandas and explained how to distinguish between happy and unhappy pandas. He was convinced that, with the freedom to move between their rooms and the open spaces of their enclosures, pandas could express their emotions. By tuning into the pandas' emotional states, Yang Tao expressed his desire to grow up alongside them as their guardian. It was clear that Yang Tao's attachment to the pandas was deep and strong, as he had cultivated a personal bond with the animals.

Love or conserve?

The two narratives thematizing fans' attachment to the giant panda – *being there* and the *reciprocal perception* – describe experiences that Morton (2021) termed the "symbiotic being", an ecological embodiment with other biological beings both inside and outside. Morton (2021) notes:

> The problem with ecological awareness and action isn't that it's horribly difficult. It's that it's too easy. You are breathing air, your bacterial microbiome is humming away, evolution is silently unfolding in the background. Somewhere, a bird is singing and clouds pass overhead. You stop reading this book and look around you . . . You don't have to be ecological. Because you are ecological.

When I journeyed into the panda fandom, I searched for a secret remedy that afforded the fans a consistent interest and love for the animal. I believed that fans were, perhaps, singled out for their connection with a more-than-human species, and they were unique in their place to have experienced this link, the resonance between a remarkable animal and the human world. How could the fans appreciate and enjoy such a beautiful connection while I had little or no knowledge about it?

Morton's words and the stories of panda fans have answers to my question: The secret of the panda love was that it was "too easy". The secret was that there was no horrifically difficult effort to obtain panda love, but that, as human beings, we are inherently driven to experience and feel it. Panda fans had no secret formula with which they built their interest in the animal, but instead an openness and willingness to look around through and with the giant panda. They are, in the words of Morton, simply being and experiencing the ecological.

What I am suggesting is that this panda love, because of its simple, easy, and ordinary form, does not mean that the conservation programs and wildlife protection projects, which often extend across locations and communities, that we have strived to construct over the years, have become worthless or pointless. Rather, the panda love, coming from a wide spectrum of people who have no formal education or training in conservation, provides a means to rethink some of the problems in the ongoing conservation agenda. The questions are: What does conservation mean if it builds on the panda love fans have fostered? Could pandas enjoy a different relationship with us through the panda love? What would giant panda conservation look like if pandas are simply "being there" and reciprocating?

Despite Yang Tao's devotion to being the panda's guardian, his reluctance to conceive this relationship through the idea of giant panda welfare and conservation was apparent.

YG: Hmm . . . As a fan, do you pay attention to issues related to the welfare and protection of giant pandas?
YT: Hmm . . . I have paid attention before. But . . .
YG: Paid attention before, so you don't pay attention anymore?
YT: Not exactly, not . . . not paying attention. Just not overly focused on it.
YG: Why not focusing on it?

YT: Because I feel that professional matters should be handled by professionals. Fans discussing these things don't have . . . much significant impact.
YG: So what's the scope of your fan activities? Your range of activities?
YT: Mainly just watching . . .
YG: What do you do as a fan?
YT: Mainly just watching . . . watching the giant pandas, and then there are some interesting activities. Mainly . . . just these surface-level activities. We don't really delve into more serious matters. We do raise suggestions when we see things . . . in the public areas.
YG: You give suggestions.

Yang Tao suggested that he had engaged with giant panda welfare and protection, a form of activism or conservation activity, before choosing to pay less attention to panda conservation. If Yang Tao's love and devotion for the giant panda had only deepened over the past 7 years, why would he suspend what appeared to be most beneficial to the animals? He explained that professionals or experts with the proper knowledge, skills, and techniques were better equipped than he was in terms of taking care of the pandas. Moreover, he pointed out that panda fans' love for the animal – perhaps precisely because it was too easy, quotidian, and ordinary – could not make much significant impact. This reflects what researchers have famously observed in the gap between conservation professionals and the public (Arlettaz et al., 2010; Roche et al., 2022; Villamagna et al., 2017), which conservation education programs must attempt to address. But does the panda love have to be professional? Do people need to receive training or education to be an expert in panda love? If panda fans have already been ecological when they come to watch their beloved animals every day, how do we reconcile this with conservation programs targeting the giant panda ecology?

Huang Hua (F, 34, junior fan) specified the difference between fans and professionals:

> As a professional panda keeper and caretaker, they undergo extensive professional training over a long period. We, as ordinary people, cannot achieve this level of expertise. Therefore, we can only participate as regular visitors, touring and observing. Other than that, we can't do much; we can only perform these basic actions.

This knowledge gap between professionals and fans highlights the limited roles fans believe they can play in contributing to the animals' well-being. In our further discussion of giant panda welfare and conservation, Huang Hua maintained her view:

YG: As a panda fan, do you pay special attention to issues regarding panda welfare and protection?

HH: Um . . . welfare, I haven't really paid attention to that. Um . . . issues, what specific issues? I haven't paid much attention to those either. I'm more of a casual fan, so I haven't focused on these details.
YG: What do you think you can do for panda welfare or protection?
HH: Um . . . if opportunities arise in the future, I will do my best to contribute and do something.

Huang Hua's answer to the question of fans' contribution to giant panda welfare and conservation was typical among my interviewees. Fans often marginalized themselves based on their perceived lack of expertise in the field of giant panda conservation, which they believed to be tasks for professionals and experts. Huang Hua categorized herself as a "casual fan", whose experience and love for the giant panda felt less significant when compared to the extensively trained keepers and caretakers. This difference in expertise and knowledge led Huang Hua to adopt a passive approach toward panda welfare and protection. While she loved and supported the giant panda, she did not view panda conservation and welfare as the primary way in which she could "give back" to the animal.

This gap between conservation professionals and panda fans persisted throughout my interviews. Two fans mentioned giant panda conservation volunteer programs but neither were active participants. More fans reflected on their limited influence on giant panda conservation and welfare, emphasizing that as mere visitors, the responsibility lay with more authoritative figures, such as the government, enterprises, or professionals. Fans suggested that communicating with others, sharing through social media, and making suggestions about giant panda welfare were ways in which they could show care for the animal. Despite their acknowledgment of the knowledge gap, Ma Juan (F, 40, dedicated fan) told me how the fan community leveraged its knowledge about visitors to reduce the noise at the giant panda enclosures:

MJ: And then, for example, when we post videos . . . For instance, sometimes those people shout loudly, we will write that in our videos.
YG: This is publicity . . . promoting pandas . . .
MJ: People are not necessarily malicious, some do indeed really not know, you know. Like, we can tell visitors try not to disturb giant pandas in their own lives. Try not to . . . she has her own life. Just visit with our eyes, not our mouths, right?
YG: Yes. My experience tells me the order at the enclosure is improving.
MJ: When you . . . when you . . . when you came to this place before, it was very noisy. You hear that sound. When visitors came to Hua Hua, they just called out for her and so on. In fact, visitors just did not understand. When they call Hua Hua, she won't come out anyway. In this case, you won't see the panda. Now people started to understand this. It was understandable that when people see Hua Hua, they get excited and enthusiastic, so they shout at her. If not shouting out at her, people also want Hua Hua to come over.

YG: Right.
MJ: Now when we are visiting Hua Hua and meeting someone calling the panda next to us, we will also stop him or her.
YG: OK.
MJ: Yes. We did these. We would tell them the panda's lifestyle and her habit.
YG: Great.
MJ: For instance, there was a day when a literary society visited the giant panda. The society was for kids, you see that . . . it was one . . . over a thousand little sparrows . . . Things could have gone out of control . . . But some kids behaved really good and we also took videos of them. Under the teacher's instruction, the kids were all quiet. This . . . this is very good. Of course there was some exceptions but I think it was because it was a really young kid.
YG: Then you should have experienced the transformation of your visiting experience, from very noisy to very quiet.
MJ: Right.
YG: How did this happen?
MJ: It's like, it happened, so many people said not to be noisy including us. You shouldn't be noisy. Because . . . because pandas can't hear our voices, and don't know. Our voices can scare the pandas, right? I see the Panda Base have reminders here, telling people to be quiet. We gradually came to this understanding that the panda likes a quiet life.
YG: Right.
MJ: For example, Hua Hua, she has been here for a long time. She is also very smart. If she feels it is too noisy, she will find a place to hide, right? In the end, as a visitor, you won't be able to see Hua Hua. If you cannot see the panda, why are you here then?
YG: Right. So don't make noise.
MJ: You don't want to take a glimpse of the giant panda's shadow, do you? You will regret it.

Ma Juan experienced a significant transformation in the atmosphere of the panda enclosure – from a noisy, chaotic environment to a much quieter, more peaceful one. My fieldwork allowed me to reciprocate Ma Juan's appreciation, as I too noticed the difference: The once loud, bustling crowds were now replaced by visitors who whispered to each other as they observed the pandas. Ma Juan's story illustrates how fans worked to improve the welfare of the giant panda through their position as tourists, adopting a philosophy of "just letting pandas be".

This internal effort – fans establishing a sense of order for other tourists – was rooted in Ma Juan's understanding of the visitors' behavior and their expectations. It was a collective endeavor that ultimately succeeded in transforming social norms, beginning with the fans' love for the animal.

Biopowered

In *The History of Sexuality*, Foucault (1990, pp. 142–143) explains biopower in these words:

> Power would no longer be dealing simply with legal subjects over whom the ultimate dominion was death, but with living beings, and the mastery it would be able to exercise over them would have to be applied at the level of life itself; it was the taking charge of life, more than the threat of death, that gave power its access even to the body.

Foucault illustrates that biopower is a modern technique designed to control and manipulate the bodies of a population. By harnessing the biological processes of the body, Foucault argues, biopower benefits the economy and efficiency of systems that maximize and anatomize the body's forces and potentials. It manages and regulates populations, focusing on the mechanisms of life such as birth, health, mortality, and longevity. In contemporary society, biopower's central goal is to "make live".

According to Foucault (1990, pp. 151–152), biopower concerns both access to and knowledge of bodily processes, and successfully integrating bodies into society:

> In any case, the purpose of the present study is in fact to show how deployments of power are directly connected to the body—to bodies, functions, physiological processes, sensations, and pleasures; far from the body having to be effaced, what is needed is to make it visible through an analysis in which the biological and the historical are not consecutive to one an other, as in the evolutionism of the first sociologists, but are bound together in an increasingly complex fashion in accordance with the development of the modern technologies of power that take life as their objective.

Biopower, in this light, means purposefully influencing and regulating bodies for the benefit of societal structure and mechanism. The power of biopower is closely tied to its understanding of bodily functions, processes, sensations, and pleasures. These bodily experiences and feelings form the foundation on which biopower transforms the population, integrating individuals into the circulation of the economy. For Foucault, human bodies – due to their vast experiential dimensions – can be accessed and exploited for larger societal purposes beyond individual understanding. Thus, biopower in contemporary society manifests as "domination, exploitation, expropriation, and, in some cases, elimination of the vital existence of some or all subjects over whom it is exercised" (Rabinow & Rose, 2006, p. 198).

Interestingly, while Foucault's concerns show that biopower is expanding its control over bodies due to the body's massive capacity to reciprocate sensations and pleasures, this accessible, responsive, and dynamic body often appears to be numbed or closed off to other external powers, such as more-than-human worlds. In *The spell of the sensuous perception*, Abram (2012, p. 47) highlights that Western civilization has created bodies that are "stifled within mass-produced structures closed off from the rest of the earth, imprisoned within technologies that plunder the living land". Abram suggests that the bodies shaped by societal structures appear "sadly superfluous and dull when we identify with our bodies and taste the world with our animal sense" (Abram, 2012, p. 47).

Here, Abram's notion focuses on bodies that are largely inaccessible to nonhuman beings. The bodies Abram refers to seem to be the result of the fine regulation and discipline imposed by biopower. Abram's observation enhances Foucault's biopower by suggesting that contemporary bodies, although subjected to powerful control, often lack the ability to respond to or understand other forms of power. The body, when controlled and regulated by biopower, loses its capacity to engage with the broader world in a more holistic, reciprocal manner.

This explanation of the consequential link between Foucault's and Abram's views on the body, however, complicates Foucault's use of the biological body as the foundation for the development of biopower. The manipulation and regulation of bodies in biopower begin with the bodies' openness to their biological dimensions and responses. For bodies to be managed effectively, they must be sensuous and reactive beings that do not become numb to external stimuli. A lively body provides both the means for and the end of biopower techniques. If "making live" is the central aim of biopower, targeting the expansion of its impact through the sensuous and pleasurable aspects of the body, how can we argue that bodies in the contemporary world are indifferent or inaccessible to more-than-human influences?

The issue, then, is not that bodies are coerced into becoming numb or deaf to their senses and experiences, but that they have been regulated and controlled to be sensible and responsive to specific forms of power or influence. In fact, we could argue that contemporary bodies, because of the meticulous and ongoing exploitation of biopower, are the most sensuous and accessible they have ever been in history. As biopower exerts its control and understanding of the body, it enhances the body's senses and pleasures through its techniques.

The question, therefore, is no longer about the human body's exclusion from the ecological world but rather about its ability to become "biopowered" by forces and powers that are more-than-human. To emerge with the nonhuman other requires a choice – an ethical decision that allows the body to be influenced by powers different from those driving societal or economic goals. These powers may include "useless" emotions, as Morton (2021) describes, or "far more expressive and diverse" relations, as Abram (2012, p. 47) states. This process involves affirming the possibility of becoming biopowered by different sources of power and by more-than-human others. It is through this process – being manipulated, lost, controlled, enchanted, taken over, enveloped, and dazed by the other – that the possibility of living in a more-than-human world can become part of daily life.

Foucault (1990) shows that human bodies are fully capable of becoming biopowered on a population scale, as people are integrated into societal structures and mechanisms that generate capital, profit, order, and discipline. As biopower consciously manipulates and controls individual bodies and their senses, ordering them into a manageable human population, it hints at how easily and vulnerably human bodies can be shaped by external influences. As Abram (2012, p. 38) writes about this possibility of creation through the body:

> Considered phenomenologically – that is, as we actually experience and live it – the body is a creative, shape-shifting entity. Certainly, it has its finite character and style, its unique textures and temperaments that distinguish it from other bodies; yet these mortal limits in no way close me off from the things around me or render my relations to them wholly predictable and determinate. On the contrary, my finite bodily presence alone is what enables me to freely engage the things around me, to choose to affiliate with certain persons or places, to insinuate myself in other lives. Far from restricting my access to things and to the world, the body is my very means of entering into relation with all things.

The resistance of a biopowered body and individuality in contemporary society, in this light, do not arise from an idealized or ideological transcendence of the spirit, mind, or soul. Instead, resistance begins where biopower itself ends – by acknowledging the freedom and possibility of becoming biopowered through an active engagement with the things around us: The panda, the animals, the plants, the more-than-human, and the other. While Foucault's conclusion about biopower demonstrates that bodies are the material targets of forces that shape populations – resources that capitalist relations exploit and invest in – choosing to become a biopowered body through engagement with the other offers a space where we can imagine, live, and appreciate the more-than-human worlds.

This resistance, then, is not an escape from biopower but a recognition of the way the body can actively engage with forces that transcend societal and economic purposes. In this context, biopower is not only about the domination and regulation of bodies for economic productivity but can also be reimagined as an embodied experience that engages with a broader, more ecological understanding of existence. By acknowledging the body's capacity to respond to, interact with, and be influenced by the more-than-human world, we create space for the possibility of more equitable and harmonious relations with the nonhuman world.

CONCLUSION

From panda fandom to crustacean fandom

When I finished my seminar at Oulu University on panda fandom, a biologist from the Biodiverse Anthropocene program approached me and expressed his wish for a crustacean fandom that could significantly increase public awareness and interest in these animals. In our discussion, we concluded that a crustacean fandom could also be an important biodiverse project. The difference between a panda fandom and a crustacean fandom, however, is that we first need to create the crustacean fandom. If the connection between panda fans and pandas is rooted in a secret formula of love or passion that I aimed to uncover in this book, we could argue that this formula might be useful in establishing a crustacean fandom. However, I have yet to crack a formula for panda fandom that could directly inform the creation of a crustacean fandom. The panda fandom is a complex phenomenon that has emerged from historical materialities and contingencies, where biopower forces and biopowered bodies intersect. Additionally, the panda fandom, to some extent, represents a more-than-human world as it engages with social and economic relations beyond the individual human subject, incorporating the giant panda and its presence within these existing relations. From an ecological perspective, the panda fandom exemplifies a space where a biopowered more-than-human world interacts with existing societal structures and relations. However, the panda fandom also maintains the power structures of these social and economic relations. While the ideologies of nation and the biopower of cuteness have shaped the creation and identity of the Chinese panda fan population, I also argue that the potential of being biopowered could offer a different, less violent way to reconnect with the nonhuman other.

So, how do we conceive of a crustacean fandom? What are the implications from the panda fandom for the crustacean fandom? How might the relationships among panda fandom, nation, capitalism, and biopower inform the creation of a crustacean fandom?

DOI: 10.4324/9781003511229-11

The first step in conceiving a crustacean fandom, it seems, is aligning the life of crustaceans with the biopower of cuteness. In Chapter 4, I discussed how the aesthetic of cuteness mobilizes biopower techniques to create lives and species through scientific knowledge about human beings. To make crustaceans attractive, it is understandable that biopowers, such as cuteness, would need to exercise their influence over these animals. In his study on cuteness, Genosko (2005) demonstrates how artists aestheticize lobsters – animals not typically seen as cute – by highlighting features like large eyes and delicate, round bodies. Genosko (2005) suggests that artists emphasize the lobster's young, small features, portraying it as a "baby lobster". Rosa et al. (2020) introduced crustacean-inspired Pokémon, called Pokécrustacea, and explored the important role crustaceans play in the Pokémon franchise. The researchers showed that the Pokécrustaceas were either inspired by real animals or combinations of different crustacean species, with certain biological features creatively addressed by the animation team. In a side-by-side comparison of German zoologist Ernst Haeckel's 1904 work *Kunstformen der Natur* and the Pokécrustaceas, the animated versions of the animals visibly targeted the effect of cuteness: The crustaceans' outlines were rounded, their eyes enlarged and prominent on bigger heads, and their bodies chubbier. While Rosa et al. (2020) believed the Pokécrustacea positively influences biodiverse consciousness among younger generations, creating and animating cute crustaceans remains an essential first step toward exploring the possibility of a crustacean fandom.

Despite cuteness as a biopower tool helping transform crustaceans into cute species that potential fans can build relationships with through this "making live" effort, I believe it is the crustacean's lack of significance and relevance to the idea of nation that jeopardizes the animals' rise to prominence within the national/international population. While the panda fandom, as demonstrated in Chapter 5, is fueled by nationalistic passions and interests and is integral to the national identity of the Chinese, it is this proximity to the nation – perhaps the most prominent institution in the modern age – that contributes to the panda's powerful impact and influence over populations managed and regulated by nations. The invention of "panda diplomacy", which capitalized on the exchange of the animal across nations, illustrates how national institutions facilitate the animal's exploitation and contribute to the creation of its trading market. The ability of crustaceans to attract attention from nations is limited, given their restricted influence on populations' identity. For example, Lewis (1989) considers the Maine lobster a regional icon, whose regional characteristics co-present with the "down east" accent, Maine humor, and the distinctive architecture of the region. Guerrero (2019) explains that the Maryland crab has a sociocultural and environmental relationship within the regional community. Unlike the panda's broad appeal through her cuteness and connection to the Chinese, crustaceans tend to be celebrated and valued within specific local regions that might have developed different or challenging relationships with nations. This does not mean that the giant panda is not also a local icon, but as Chapter 1 shows, local interest in the giant panda was an affirmation of national unity in response to the Western colonial and imperialist invasions since the discovery of the animal.

The diversity and beauty of crustaceans have made them prominent in the public arena, as they serve as food and, more recently, as pet food and entertainment (Galil et al., 2011). Galil et al. (2011) point out that our knowledge of marine crustaceans has been insufficient to fully understand their history, diversity, distribution, and impacts. In contrast to the crustaceans' diversity, global distribution, and large population, the giant panda's rarity and geographical specificity add to its appeal to the human population. While panda fans have been able to build emotional attachments to the fluffy animal, can fans establish a similar bond with crustaceans – animals we often consume as food? Evans and Miele (2012, p. 300) have used the term "foodsensing" to describe the "hybrid process through which consumers simultaneously sense and make sense of food". Grounded in embodied sensations, where food consumed merges with and becomes part of our bodies, Evans and Miele (2012, p. 301) suggest that eating, or the act of consuming, is a "lively process" that creatively makes novel "realities" or "matters" through the integration of the bodies. In this foodsensing space, Evans and Miele resonate with Abram (2012), showing that sensing has always already been "making sense", through which new bodies emerge and new orders are created.

In this light, foodsensing also leads to an experience of "being biopowered", as we allow food to become part of our bodies, integrating with the flesh we live through every day. This physical intimacy between human bodies and the crustaceans consumed as food underscores an embodied and biopowered existence that has always been more-than-human. However, one significant difference between the biopowered experience with the giant panda and the crustacean is that panda fans can, to a certain degree, perceive the panda's happiness and contentment despite its confinement. As panda fans, they experience the panda's loving and happy presence merging with their own bodies. How, then, could we imagine happily consumed crustaceans while taking a bite? Do crustaceans willingly sacrifice their lives for the well-being and benefit of human bodies and health? If the panda's playful happiness can, for panda fans, biopower their intimacy and experience with the animal, how should we make sense of the emotions and feelings of crustaceans being eaten? Should we become vulnerable and open to the experiences of being consumed? The impossibility and difficulty of imagining happy and content crustaceans being consumed make foodsensing an anthropocentric violence that hinders the biopowered experience. I think this prohibits the conception and construction of a crustacean fandom. Creating a crustacean fandom that allows the animals to "just be there", reciprocating with humans, seems to disturb the ongoing consumption of the species. Having crustaceans just be there, loved by their fans, could challenge the existing order of eating and consuming the animals. Reciprocal participation with the crustaceans could open up novel and unexpected ways of making sense and sensing through human bodies living in the more-than-human world.

This potential for being biopowered, as exemplified by the panda fandom, illuminates the evolving human–animal entanglements alongside the development of the human population and its future. When I asked panda fans where this fandom and their love for pandas would lead them in 3 years, I anticipated responses that would provide insight. However, fans expressed vagueness and uncertainty, or admitted they had not previously considered the question, despite believing their affection for the animal would persist and deepen over time. Could this enduring love for pandas be strong enough to shape a future for both the fans and the animal? What expectations did I have for this panda love? Was I envisioning a scenario where fans and pandas could form legally recognized, happily-ever-after unions akin to contemporary society's legitimization and protection of human marriage? What possibilities or agendas could we imagine that would lend more credence to these more-than-human entanglements? Perhaps we do not need a fan–panda marriage but rather a panda fandom that continues to integrate animals and other species, whether deemed cute or not, into our human experiences and perceptions. We might call it a more-than-human fandom, where pandas, humans, other animals, and beings on this planet are free to engage and coexist through making sense.

BIBLIOGRAPHY

Abram, D. (2012). *The spell of the sensuous: Perception and language in a more-than-human world*. Vintage.

Aden, R. C., Armfield, G. G., Beard, D. E., Berg, K., Billings, A. C., Boone, J., Bowman, N. D., Brady, K., Chidester, P. J., & Fingerhut, D. (2011). *Sports fans, identity, and socialization: Exploring the fandemonium*. Lexington Books.

Agamben, G., & Sacer, H. (1995). Sovereign power and bare life. *Homo Sacer*, *1*.

Agrawal, A., Gupta, A., & Yousaf, A. (2018). Like it but do not comment: Manipulating the engagement of sports fans in social media. *International Journal of Sport Management and Marketing*, *18*(4), 340–356.

Ahmed, S. (2004). Affective economies. *Social Text*, *22*(2), 117–139.

Alan Fine, G., & Hallett, T. (2014). Stranger and stranger: Creating theory through ethnographic distance and authority. *Journal of Organizational Ethnography*, *3*(2), 188–203.

Aleteia. (2023). *How a French catholic priest "discovered" the giant panda*. Retrieved March20,2023,fromhttps://aleteia.org/2017/08/16/how-a-french-catholic-priest-discovered-the-giant-panda/

Allison, A. (2003). Portable monsters and commodity cuteness: Poke´mon as Japan's new global power. *Postcolonial Studies*, *6*(3), 381–395.

Althusser, L. (2006). Ideology and ideological state apparatuses (notes towards an investigation). *The Anthropology of the State: A Reader*, *9*(1), 86–98.

Anderlini, J. (2017). Panda politics: The hard truth about China's cuddliest diplomat. *FT Magazine*. Retrieved August 18, 2022, from https://www.ft.com/content/8a04a532-be92–11e7–9836-b25f8adaa111

Anderson, B. (2012). Affect and biopower: Towards a politics of life. *Transactions of the Institute of British Geographers*, *37*(1), 28–43.

Angier, N. (2006). The cute factor. *The New York Times*, F1 (L).

Arlettaz, R., Schaub, M., Fournier, J., Reichlin, T. S., Sierro, A., Watson, J. E., & Braunisch, V. (2010). From publications to public actions: When conservation biologists bridge the gap between research and implementation. *BioScience*, *60*(10), 835–842.

Bamshad, M. J., & Olson, S. E. (2003). Does race exist? *Scientific American*, *289*(6), 78–85.

Barstow, G. (2019). On the moral standing of animals in Tibetan Buddhism. *Études mongoles et sibériennes, centrasiatiques et tibétaines*, (50).

Barua, M. (2016). Lively commodities and encounter value. *Environment and Planning D*, *34*(4), 725–744.

Barua, M. (2020). Affective economies, pandas, and the atmospheric politics of lively capital. *Transactions of the Institute of British Geographers*, *45*(3), 678–692.

Bennett, T. (2013). *The birth of the museum: History, theory, politics*. Routledge.

Bentham, J. (2020). *The panopticon writings*. Verso Books.

Berger, J. (2015). *About looking*. Bloomsbury Publishing.

Bhabha, H. (2015). *Debating cultural hybridity: Multicultural identities and the politics of anti-racism*. Bloomsbury Publishing.

Bhagwat, S. A., Kushalappa, C. G., Williams, P. H., & Brown, N. D. (2005). The role of informal protected areas in maintaining biodiversity in the Western Ghats of India. *Ecology and Society*, *10*(1).

Bhattacharya, K., Ghosh, A., Monsivais, D., Dunbar, R., & Kaski, K. (2017). Absence makes the heart grow fonder: Social compensation when failure to interact risks weakening a relationship. *EPJ Data Science*, *6*(1), 1–10.

Blewitt, J. (2013). What's new pussycat? A genealogy of animal celebrity. *Celebrity Studies*, *4*(3), 325–338.

Bo, A. (2003). 再谈熊猫之称谓. 文史杂志, *2*, 74.

Booth, P. J. (2015). Fandom: The classroom of the future. *Transformative Works and Cultures*, *19*(0).

Braidotti, R. (2013). *The posthuman*. John Wiley & Sons.

Brambell, M. (1976). The giant panda (*Ailuropoda melanoleuca*). *The Transactions of the Zoological Society of London*, *33*(2), 85–92.

Brenkert, G. G. (1979). Freedom and private property in Marx. *Philosophy & Public Affairs*, 122–147.

Brown, K. W., & Ryan, R. M. (2003). The benefits of being present: Mindfulness and its role in psychological well-being. *Journal of Personality and Social Psychology*, *84*(4), 822.

Buckingham, K. C., David, J. N. W., & Jepson, P. (2013). Environmental reviews and case studies: Diplomats and refugees: Panda diplomacy, soft "cuddly" power, and the new trajectory in panda conservation. *Environmental Practice*, *15*(3), 262–270.

Busse, K. (2015). Fan labor and feminism: Capitalizing on the fannish labor of love. *Cinema Journal*, *54*(3), 110–115.

Butler, J. (2002). *Gender trouble*. Routledge.

Callahan, W. A. (2004). National insecurities: Humiliation, salvation, and Chinese nationalism. *Alternatives*, *29*(2), 199–218.

Callahan, W. A. (2009). *China: The pessoptimist nation*. OUP.

Cantor, L. (2020). Zhuangzi on "happy fish" and the limits of human knowledge. *British Journal for the History of Philosophy*, *28*(2), 216–230.

Chen, Y., Lehto, X. Y., & Cai, L. (2013). Vacation and well-being: A study of Chinese tourists. *Annals of Tourism Research*, *42*, 284–310.

Cheng, C. (2023, April 28, 2023). China welcomes home Ya Ya, the not-so-fluffy panda, amid acrimony and accusations. *NBC News*. https://www.nbcnews.com/news/world/panda-yaya-china-memphis-zoo-diplomacy-washington-beijing-geopolitics-rcna81566

Chiang, Y. (1945). *The story of Ming*. Puffin.

China, S. M. o. (2023). *Western naming of giant panda in recent times*. Retrieved March 20, 2023, from http://www.kepu.net.cn/english/giantpanda/giantpanda_know/200409230030.html

Chinoy, M. (1975). Pandamonium, or everything you always wanted to know about pandas. *New China, Spring*, 15–17.

Cho, S., Dydynski, J. M., & Kang, C. (2022). Universality and specificity of the Kindchenschema: A cross-cultural study on cute rectangles. *Psychology of Aesthetics, Creativity, and the Arts*, *16*(4), 719.

Chow, R. (1995). *Primitive passions: Visuality, sexuality, ethnography, and contemporary Chinese cinema*. Columbia University Press.

Chow, R. (1996). Women in the Holocene: Ethnicity, fantasy, and the film the Joy Luck Club. In C. Luke (Ed.), *Feminisms and Pedagogies of Everyday Life* (pp. 204–221). Albany, NY: State University of New York Press.

Chow, R. (2002). *The protestant ethnic and the spirit of capitalism*. Columbia University Press.
Chow, R. (2003). *Woman and Chinese modernity: The politics of reading between West and East* (Vol. 75). University of Minnesota Press.
Christou, P. A. (2020). *Philosophies of hospitality and tourism: Giving and receiving* (Vol. 91). Channel View Publications.
Chuang, J. (2020). *Beneath the China boom: Labor, citizenship, and the making of a rural land market*. University of California Press.
Clark, A. (1998). *Being there: Putting brain, body, and world together again*. MIT Press.
Cohen, E. (1988). Authenticity and commoditization in tourism. *Annals of Tourism Research, 15*(3), 371–386.
Cohen, E. (2010). Panda and elephant–contesting animal icons in Thai tourism. *Journal of Tourism and Cultural Change, 8*(3), 154–171.
Cohen, P. A. (2002). Remembering and forgetting national humiliation in twentieth-century China. *Twentieth-Century China, 27*(2), 1–39.
Collard, R.-C. (2020). *Animal traffic: Lively capital in the global exotic pet trade*. Duke University Press.
Colléony, A., Clayton, S., Couvet, D., Saint Jalme, M., & Prévot, A.-C. (2017). Human preferences for species conservation: Animal charisma trumps endangered status. *Biological Conservation, 206*, 263–269.
Cooper, M. (2013). The strategy of default: Liquid foundations in the house of finance. *Polygraph: An International Journal of Culture and Politics, 23*(24), 79–96.
Cooper, M. E. (2011). *Life as surplus: Biotechnology and capitalism in the neoliberal era*. University of Washington Press.
Cotter, J. (2016). The new class common-sense. In J. Cotter, K. DeFazio, R. Faivre, A. Sahay, J. P. Torrant, S. Tumino, & R. Wilkie (Eds.), *Human, all too (post) human: The humanities after humanism*. Rowman & Littlefield.
Cox, L. (2020). *Nationalism: Themes, theories, and controversies*. Springer.
Croke, V. (2009). *The lady and the panda: The true adventures of the first American explorer to bring back China's most exotic animal*. Random House.
Cui, N., Wu, T., Wang, Y.-C., Zou, H., Axmacher, J. C., Sang, W., & Guo, L. (2022). Buddhist monasteries facilitated landscape conservation on the Qinghai–Tibetan Plateau. *Landscape Ecology, 37*(6), 1559–1572.
Dajches, L. (2022). Finding the self through others: Exploring fandom, identification, and self-concept clarity among US adolescents. *Journal of Children and Media, 16*(1), 107–116.
Dale, J. P. (2017). The appeal of the cute object. In J. P. Dale, J. Goggin, J. Leyda, A. P. McIntyre, & D. Negra (Eds.), *The aesthetics and affects of cuteness* (pp. 35–55). Routledge.
Dale, J. P. (2023). *Irresistible: How cuteness wired our brains and conquered the world*. Profile Books.
Dale, J. P., Goggin, J., Leyda, J., McIntyre, A. P., & Negra, D. (2016). *The aesthetics and affects of cuteness*. Routledge.
Darwin, C. (1877). A biographical sketch of an infant. *Mind, 2*(7), 285–294.
Dawkins, R. (1996). *The blind watchmaker: Why the evidence of evolution reveals a universe without design*. WW Norton & Company.
De Bloom, J., Geurts, S. A., Taris, T. W., Sonnentag, S., de Weerth, C., & Kompier, M. A. (2010). Effects of vacation from work on health and well-being: Lots of fun, quickly gone. *Work & Stress, 24*(2), 196–216.
De Kosnik, A. (2012). Fandom as free labor. In *Digital labor* (pp. 98–111). Routledge.
De Lauretis, T. (1984). *Alice doesn't: Feminism, semiotics, cinema* (Vol. 316). Indiana University Press.
de Vere, A. J., & Kuczaj, S. A. (2016). Where are we in the study of animal emotions? *Wiley Interdisciplinary Reviews: Cognitive Science, 7*(5), 354–362.
de Waal, F. B., & Andrews, K. (2022). The question of animal emotions. *Science, 375*(6587), 1351–1352.

Doebel, S., Stucke, N. J., & Pang, S. (2022). Kindchenschema and cuteness elicit interest in caring for and playing with young children, but less so when children are masked. *Scientific Reports*, *12*(1), 11903.

Dreyfus, H. L. (1990). *Being-in-the-world: A commentary on Heidegger's being in time, division I* (Vol. 1). MIT Press.

Duffett, M. (2013). *Understanding fandom: An introduction to the study of media fan culture*. Bloomsbury Publishing.

Dydynski, J. M. (2020). Modeling cuteness: Moving towards a biosemiotic model for understanding the perception of cuteness and kindchenschema. *Biosemiotics*, *13*(2), 223–240.

Emel, J., & Wolch, J. (1998). Witnessing the animal moment. In J. Wolch & J. Emel (Eds.), *Animal Geographies* (pp. 1–24). London: Verso.

Evans, A. B., & Miele, M. (2012). Between food and flesh: How animals are made to matter (and not matter) within food consumption practices. *Environment and Planning D: Society and Space*, *30*(2), 298–314.

Fanon, F. (1970). *Black skin, white masks. The future*. Paladin.

Fejes, A., & Dahlstedt, M. (2013). *The confessing society: Foucault, confession and practices of lifelong learning*. Routledge.

Feldshuh, H. (2018). Gender, media, and myth-making: Constructing China's leftover women. *Asian Journal of Communication*, *28*(1), 38–54.

Feng, W., Xu, Y., Huang, H., & Wang, T. (2022). Kawai vs. Whimsical: The influence of cuteness types of luxury brands on consumers' preferences. *Acta Psychologica Sinica*, *54*(3), 313.

Ferris, K. O., & Ferris, K. O. (2011). Building characters: The work of celebrity impersonators. *Journal of Popular Culture*, *44*(6).

Foucault, M. (1988). An aesthetics of existence. In. L. D. Kritzman (Ed.), *Politics, philosophy, culture: Interviews and other writings, 1977–1984* (pp. 47–53). New York: Routledge.

Foucault, M. (1977b). *Discipline and punish: The birth of the prison*. Penguin.

Foucault, M. (1990). *The history of sexuality: An introduction* (Robert Hurley, Trans.; Vol. I, p. 95, pp. 1–160). Vintage.

Foucault, M. (2003). *"Society must be defended": Lectures at the Collège de France, 1975–1976* (Vol. 1). Macmillan.

Galil, B. S., Clark, P. F., & Carlton, J. T. (2011). *In the wrong place – alien marine crustaceans: Distribution, biology and impacts* (Vol. 6). Springer Science & Business Media.

Ganesko, G. (2005). Natures and cultures of cuteness. *Invisible Culture: An Electronic Journal for Visual Culture*, 9(Fall), 1–19. https://www.rochester.edu/in_visible_culture/Issue_9/genosko.html

Gellner, E. (2008). *Nations and nationalism*. Cornell University Press.

Giles, D. C. (2013). Animal celebrities. *Celebrity Studies*, *4*(2), 115–128.

Goodman, M. K., & Barnes, C. (2011). Star/poverty space: The making of the 'development celebrity'. *Celebrity Studies*, *2*(1), 69–85.

Gray, J., Sandvoss, C., & Harrington, C. L. (2017). *Fandom: Identities and communities in a mediated world*. NYU Press.

Gruenfeld, D. H., Inesi, M. E., Magee, J. C., & Galinsky, A. D. (2008). Power and the objectification of social targets. *Journal of Personality and Social Psychology*, *95*(1), 111.

Guerrero, P. (2019). *The Smith Island crab coop: Women's work in the Chesapeake Bay*. Indiana University.

Guo, Y., & Fennell, D. (2023). What makes the giant panda a celebrity? *Celebrity Studies*, 1–16.

Gupta, A. (2022). Panda power? Chinese soft power in the Era of COVID–19. *Prism*, *10*(1), 40–56.

Haahtela, T. (2019). A biodiversity hypothesis. *Allergy*, *74*(8), 1445–1456.

Hall, B., & Hallgrímsson, B. (2008). *Strickberger's evolution*. Jones Bartlett Learning.

Hall, S. (1999). Un-settling 'the heritage', re-imagining the post-nation whose heritage? *Third Text*, *13*(49), 3–13.

Hamwey, M. K., Rolan, E. P., Jensen, A. C., & Whiteman, S. D. (2019). "Absence makes the heart grow fonder": A qualitative examination of sibling relationships during emerging adulthood. *Journal of Social and Personal Relationships*, *36*(8), 2487–2506.

Han, Q. (2012). The ties that bind: An overview of traditional Chinese family ethics. *Journal of Chinese Studies*, *1*(1), 85–99.

Hao, A. (2020). Understanding consumer fandom: Literature review and conceptual framework. In C. L. Wang (Ed.), *Handbook of research on the impact of fandom in society and consumerism* (pp. 18–37). IGI Global.

Haraway, D. J. (2013). *When species meet* (Vol. 3). University of Minnesota Press.

Harney, A. (2008). *The China price: The true cost of Chinese competitive advantage*. Penguin.

Harper, K. (2013). Culture, nature, and history: The case of ancient sexuality. *Comparative Studies in Society and History*, *55*(4), 986–1016.

Harrell, S. (1993). Linguistics and hegemony in China. *International Journal of the Sociology of Language, 103*, 97–114.

Harris, D. (2000). *Cute, quaint, hungry, and romantic: The aesthetics of consumerism* (Vol. 10). Basic Books.

Hartig, F. (2013). Panda diplomacy: The cutest part of China's public diplomacy. *The Hague Journal of Diplomacy*, *8*(1), 49–78.

He, Z., & Jiang, X. (1842). *Liu Ba Ting Zhi* 留坝厅志. Office of Local Chronicles of Shaanxi Province. http://dfz.shaanxi.gov.cn/sqzlk/dqcs/dacswz/hzs_16205/qwbyd_9_1/

Heater, C. (1992). Being there: The subjective experience of presence. *Presence Teleoperators Virtual Environments*, *1*(2), 262–271.

Heidegger, M. (1977). The age of the world picture. In *Science and the quest for reality* (pp. 70–88). Springer.

Heinrich, A. L. (2018). *Chinese Surplus: Biopolitical aesthetics and the medically commodified body*. Duke University Press.

Heller, C. (2015). *How America fell in love with the giant panda*. Retrieved March 20, 2023, from https://www.smithsonianmag.com/history/how-america-fell-love-giant-panda-180956692/

Heyes, C. J. (2014). Subjectivity and power. In *Michel Foucault* (pp. 159–172). Routledge.

Hills, M. (2003). *Fan cultures*. Routledge.

Ho, D. Y., Xie, W., Liang, X., & Zeng, L. (2012). Filial piety and traditional Chinese values: A study of high and mass cultures. *PsyCh Journal*, *1*(1), 40–55.

Hoffman, M. (2014). Disciplinary power. In *Michel Foucault* (pp. 27–39). Routledge.

Hollinshead, K. (1999). Surveillance of the worlds of tourism: Foucault and the eye-of-power. *Tourism Management*, *20*(1), 7–23.

Hou, G., & Jiang, Y. (1991). Jiāróng Zàngzú yǎn zhōng de zìrán měi [The perception of natural beauty among the Jiarong Tibetans]. Wén Shǐ Zázhì [*Journal of Literature and History*], (3), 27–29.

Hou, W. (1997). Reflections on Chinese traditional ideas of nature. *Environmental History*, *2*(4), 482–493.

Hu, D. (1994). 宝兴大熊猫面世. In G. Huang, M. Gui, & D. Hu (Eds.), 宝兴大熊猫. Sichuan Ya'an Regional Publishing.

Huang, G., Ping, X., Xu, W., Hu, Y., Chang, J., Swaisgood, R. R., Zhou, J., Zhan, X., Zhang, Z., & Nie, Y. (2021). Wildlife conservation and management in China: Achievements, challenges and perspectives. *National Science Review*, *8*(7), nwab042.

Huang, Z. A., & Wang, R. (2019). The new 'cat' of the internet: China's panda diplomacy on Twitter. In F. Frandsen, W. Johansen, R. Tench, & S. Romenti (Eds.), *Big ideas in public relations research and practice* (pp. 69–84). Emerald Publishing Limited.

Huang, Z. A., & Wang, R. (2020). "Panda engagement" in China's digital public diplomacy. *Asian Journal of Communication*, *30*(2), 118–140.

Huijer, M. (1999). The aesthetics of existence in the work of Michel Foucault. *Philosophy & Social Criticism*, *25*(2), 61–85.

Hussey, K. (n.d.). *Treasures from the odontological collection*. Retrieved May 14, 2024, from https://publishing.rcseng.ac.uk/doi/pdf/10.1308/147363513X13690603817427

Imbriaco, A. (2006). *The Giant panda: Help save this endangered species!* Enslow Publishers, Inc.

Jarić, I., Courchamp, F., Correia, R. A., Crowley, S. L., Essl, F., Fischer, A., González-Moreno, P., Kalinkat, G., Lambin, X., & Lenzner, B. (2020). The role of species charisma in biological invasions. *Frontiers in Ecology and the Environment*, *18*(6), 345–353.

Jenkins, H. (2006). *Convergence culture: Where old and new media collide*. New York University Press.

Jenkins, H. (2007). Afterword: the future of fandom. In J. Gray, C. Sandvoss, & C. L. Harrington (Eds.), *Fandom* (p. 357364). New York University Press.

Jenkins, H. (2012). *Textual poachers: Television fans and participatory culture*. Routledge.

Ji, A. (2023). 熊猫"饭圈化" 边界在哪里? *Xiongmao "fanquanhua" bianjia zainali?* Retrieved June 14, 2023, from https://www.163.com/dy/article/I681UGV20514BE2Q.html

Ji, Y. (2015). Between tradition and modernity: "Leftover" women in Shanghai. *Journal of Marriage and Family*, *77*(5), 1057–1073.

Jia, H. 試論明清封建皇朝的土司制及改土歸流. 四川大学学报*(*社会科学版*)*, *2*, 59–75.

Jiang, H. (2022). Science, commerce, and politics: China's giant pandas going global (1869–1948). *Journal of Modern Chinese History*, *16*(2), 206–220.

Kant, I. (1952). *Critique of Judgement* (J. Meredith, Trans.; N. Walker, Ed., 2nd ed.). Oxford University Press.

Kaufman, A. A. (2010). The "century of humiliation," then and now: Chinese perceptions of the international order. *Pacific Focus*, *25*(1), 1–33.

Keat, R. (2013). 12 The human body in social theory: Reich, Foucault and the repressive hypothesis. In *Socialism, feminism and philosophy* (pp. 277–306). Routledge.

Kirksey, S. E., & Helmreich, S. (2010). The emergence of multispecies ethnography. *Cultural Anthropology*, *25*(4), 545–576.

Knudsen, B. T., & Waade, A. M. (2010). Performative authenticity in tourism and spatial experience: Rethinking the relations between travel, place and emotion. In B. T. Knudsen & A. M. Waade (Eds.), *Re-investing authenticity: Tourism, place and emotions* (pp. 1–19). Channel View Publications.

Kotzé, D. (2002). Doing participatory ethics. In D. Kotzé, J. Myburg, J. Roux & Associates (Eds.), *Ethical ways of being* (pp. 1–34). Ethics Alive: Pretoria.

Kringelbach, M. L., Stark, E. A., Alexander, C., Bornstein, M. H., & Stein, A. (2016). On cuteness: Unlocking the parental brain and beyond. *Trends in Cognitive Sciences*, *20*(7), 545–558.

Kristeva, J. (2024). *Powers of horror: An essay on abjection*. Columbia University Press.

Krueger, D. W., & Krueger, L. N. (2018). Animals in children's stories. In *Cultural zoo* (pp. 127–143). Routledge.

LaMothe, K. L. (2015). *Why we dance: A philosophy of bodily becoming*. Columbia University Press.

Landim, A. S., de Menezes Souza, J., Dos Santos, L. B., de Freitas Lins-Neto, E. M., da Silva, D. T., & Ferreira, F. S. (2023). Food taboos and animal conservation: A systematic review on how cultural expressions influence interaction with wildlife species. *Journal of Ethnobiology and Ethnomedicine*, *19*(1), 31.

Latimer, J., & Miele, M. (2013). Naturecultures? Science, affect and the non-human. *Theory, Culture Society*, *30*(7–8), 5–31.

Lawson, C. E., & Draper, J. (2021). Working the red carpet: A framework for analysing celebrities' red carpet labour. *Celebrity Studies*, *12*(4), 635–648.

Lehmann, H., Roth, H., & Schankweiler, K. (2019). Affective economy. In *Affective societies* (pp. 140–151). Routledge.

Lewis, G. H. (1989). The Maine lobster as regional icon: Competing images over time and social class. *Food and Foodways*, *3*(4), 303–316.

Li, J., Wang, D., Yin, H., Zhaxi, D., Jiagong, Z., Schaller, G. B., Mishra, C., Mccarthy, T. M., Wang, H., & Wu, L. (2014). Role of Tibetan Buddhist monasteries in snow leopard conservation. *Conservation Biology*, *28*(1), 87–94.

Lin, H., & Lin, L. (2005). Yě tán dà xióngmāo zhī jīn gǔ chēngwèi [On the historical and contemporary names of the giant panda]. Běijīng Línyè Dàxué Xuébào: Shèhuì Kēxué Bǎn [*Journal of Beijing Forestry University: Social Science Edition*], *4*(1), 49–52.
Lindburg, D., & Baragona, K. (2004). *Giant pandas: Biology and conservation*. University of California Press.
Lipsman, A., Mudd, G., Rich, M., & Bruich, S. (2012). The power of "like": How brands reach (and influence) fans through social-media marketing. *Journal of Advertising Research*, *52*(1), 40–52.
Lorch, M., & Hellal, P. (2010). Darwin's "natural science of babies". *Journal of the History of the Neurosciences*, *19*(2), 140–157.
Lorenz, K. (1971). *Studies in animal and human behaviour* (Vol. 2). Harvard University Press.
Lorimer, J. (2015). *Wildlife in the Anthropocene: Conservation after nature*. University of Minnesota Press.
Lovell, R., Wheeler, B. W., Higgins, S. L., Irvine, K. N., & Depledge, M. H. (2014). A systematic review of the health and well-being benefits of biodiverse environments. *Journal of Toxicology and Environmental Health, Part B*, *17*(1), 1–20.
Lozano Delmar, J., Sanchez-Martin, M., & Muñiz Velázquez, J. A. (2018). To be a fan is to be happier: Using the eudaimonic spectator questionnaire to measure eudaimonic motivations in Spanish fans. *Journal of Happiness Studies*, *19*(1), 257–276.
Lv, X., Liu, Y., Luo, J., Liu, Y., & Li, C. (2021). Does a cute artificial intelligence assistant soften the blow? The impact of cuteness on customer tolerance of assistant service failure. *Annals of Tourism Research*, *87*, 103114.
Lynch, R. A. (2014). Foucault's theory of power. In *Michel Foucault* (pp. 13–26). Routledge.
Ma, K., Liu, D., Wei, R., Zhang, G., Xie, H., Huang, Y., Li, D., Zhang, H., & Xu, H. (2016). Giant panda reintroduction: Factors affecting public support. *Biodiversity and Conservation*, *25*, 2987–3004.
MacCannell, D. (1976). *The tourist: A new theory of the leisure class*. Schocken Books.
Macdonald, S. (2007). Exhibitions of power and powers of exhibition: An introduction to the politics of display. In *Museums and their communities* (pp. 176–196). Routledge.
Madon, J. (2022). Free repair against the consumer society: How repair cafés socialize people to a new relationship to objects. *Journal of Consumer Culture*, *22*(2), 534–550. https://doi.org/10.1177/1469540521990871
Magramo, K. (2023, May 29). Giant panda Ya Ya's arrival at Beijing Zoo sparks fresh outpouring of online pride. *CNN*. https://www.cnn.com/2023/05/29/asia/china-us-ya-ya-panda-returns-beijing-zoo-intl-hnk/index.html
Malhi, Y. (2017). The concept of the Anthropocene. *Annual Review of Environment and Resources*, *42*, 77–104.
Marshall, P. D. (1998). Confession and identity. *M/C: A Journal of Media and Culture*, *1*(3). Retrieved from http://journal.mediaculture.org.au/9810/conf.php
Marshall, P. D. (2014). *Celebrity and power: Fame in contemporary culture*. University of Minnesota Press.
Marx, K. (2000). *Karl Marx: Selected writings*. Oxford University Press.
Marx, K. (2019). "He's so Fluffy I'm gonna die!" Cute responses by hikers to autonomous animals on the appalachian trail. *Anthrozoös*, *32*(1), 89–101.
May, S. (2019). *The power of cute*. Princeton University Press.
May, T., & Hauser, C. (2023). A panda is coming home, and her Chinese fans say it's about time. *The New York Times*. https://www.nytimes.com/2023/04/12/world/asia/panda-china-memphis-zoo-ya-ya.html
McClintock, A. (2013). *Imperial leather: Race, gender, and sexuality in the colonial contest*. Routledge.
McInroy, L. B., & Craig, S. L. (2018). Online fandom, identity milestones, and self-identification of sexual/gender minority youth. *Journal of LGBT Youth*, *15*(3), 179–196.
Mendl, M., Neville, V., & Paul, E. S. (2022). Bridging the gap: Human emotions and animal emotions. *Affective Science*, *3*(4), 703–712.

Merish, L. (1996). Cuteness and commodity aesthetics. In R. Garland-Thomson (Ed.), *Freakery: Cultural spectacles of the extraordinary body* (pp. 185–203). New York University Press.
Michelson, B. (2023). *Watch: Update on YaYa the giant panda: Life is good in China.* Retrieved May 10, 2024, from https://www.idausa.org/campaign/wild-animals-and-habitats/latest-news/yaya-giant-panda-thriving-in-china/#:~:text=YaYa%20is%20living%20peacefully%20in,the%20need%20to%20serve%20anyone
Milner, R. M. (2009). Working for the text: Fan labor and the new organization. *International Journal of Cultural Studies, 12*(5), 491–508.
Mitchell, T. (1991). *Colonising Egypt: With a new preface*. University of California Press.
Montford, K. S., & Taylor, C. (2020). Colonialism and animality: An introduction. In *Colonialism and Animality* (pp. 1–16). Routledge.
Moore, K., Buchmann, A., Månsson, M., & Fisher, D. (2021). Authenticity in tourism theory and experience. Practically indispensable and theoretically mischievous? *Annals of Tourism Research, 89*, 103208.
Morton, T. (2021). *All art is ecological*. Penguin UK.
Mulhall, S. (2014). *On being in the world (Routledge Revivals): Wittgenstein and Heidegger on seeing aspects*. Routledge.
Mulvey, L. (2013). Visual pleasure and narrative cinema. In *Feminism and film theory* (pp. 57–68). Routledge.
Nenkov, G. Y., & Scott, M. L. (2014). "So cute I could eat it up": Priming effects of cute products on indulgent consumption. *Journal of Consumer Research, 41*(2), 326–341.
Ngai, S. (2005). The cuteness of the avant-garde. *Critical Inquiry, 31*(4), 811–847.
Ngai, S. (2010). Our aesthetic categories. *PMLA, 125*(4), 948–958.
Ngai, S. (2012). *Our aesthetic categories: Zany, cute, interesting*. Harvard University Press.
Ngai, S. (2020). *Theory of the Gimmick: Aesthetic judgment and capitalist form*. Harvard University Press.
Nicholls, H. (2012). *The way of the panda: The curious history of China's political animal*. Pegasus Books.
Nicholls, H. (2013). *Meet Ming, the panda who left China to boost Britain's wartime morale*. Retrieved March 20, 2024, from https://www.theguardian.com/science/2013/jun/13/ming-panda-pioneer-britain
Nicholls, H. (2014). The 'first' giant panda and how it ended up in Paris. *The Guardian*. https://www.theguardian.com/science/animal-magic/2014/jan/06/the-first-giant-panda-and-how-it-ended-up-in-paris
Nijhawan, S., & Mihu, A. (2020). Relations of blood: Hunting taboos and wildlife conservation in the Idu Mishmi of Northeast India. *Journal of Ethnobiology, 40*(2), 149–166.
Nunn, H., & Biressi, A. (2010). 'A trust betrayed': Celebrity and the work of emotion. *Celebrity Studies, 1*(1), 49–64.
O'Gorman, E., & Gaynor, A. (2020). More-than-human histories. *Environmental History, 25*(4), 711–735.
Okafor, L. E., Tan, C. M., & Khalid, U. (2021). Does panda diplomacy promote Chinese outbound tourism flows? *Journal of Hospitality and Tourism Management, 49*, 54–64.
Osgood, W. (1931). Roosevelts' giant panda group installed in William V. Kelley Hall. *Field Museum News,* 1. https://core.ac.uk/download/pdf/4521528.pdf
Osterhammel, J. (1986). Semi-colonialism and informal empire in twentieth-century China: Towards a framework of analysis. In W. Mommsen, & J. Osterhammel (Eds.), *Imperialism and after: continuities and discontinuities* (pp. 291–314). London: Allen & Unwin.
Pachuau, J. L., & van Schendel, W. (2022). *Entangled lives: Human–animal–plant histories of the Eastern Himalayan triangle*. Cambridge University Press.
Pacini-Ketchabaw, V., Taylor, A., & Blaise, M. (2016). Decentring the human in multispecies ethnographies. In *Posthuman research practices in education* (pp. 149–167). Springer.
Painter, C. (2016). Non-human animals within contemporary capitalism: A Marxist account of non-human animal liberation. *Capital & Class, 40*(2), 327–345.

Pallas, N. (2017). Giant panda cub makes debut in Tokyo. *Kyodo News*. Retrieved February 20, 2024, from https://english.kyodonews.net/news/2017/12/510946ae65e1-giant-panda-cub-makes-debut-in-tokyo.html

Peddie, C. (2010). *Wang Wang leaves giant panda debt*. Retrieved March 23, 2024, from https://www.couriermail.com.au/ipad/wang-wang-leaves-giant-panda-debt/news-story/0427d68edd257c15f6e7bb15bdecdb33

Piacente, A. (2020). Fans and identity. *Sport, Ethics and Philosophy*, *14*(1), 92–105.

Rabinow, P. (2022). Artificiality and enlightenment: From sociobiology to biosociality. In *The ethics of biotechnology* (pp. 101–122). Routledge.

Rabinow, P., & Rose, N. (2003) Introduction: Foucault Today. In P. Rabinow, & N. Rose (Eds.), *The Essential Foucault: Selections from Essential Works of Foucault, 1954–1984* (pp. 1–30). New York: The New Press.

Rabinow, P., & Rose, N. (2006). Biopower today. *BioSocieties*, *1*(2), 195–217.

Rahman, Z., Suberamanian, K., Zanuddin, H., Moghavvemi, S., & Nasir, M. H. N. M. (2016). Social media engagement metric analysis- "study on fan page content". *Journal of Telecommunication, Electronic and Computer Engineering (JTEC)*, *8*(8), 71–76.

Ratcliffe, M. (2005). The feeling of being. *Journal of Consciousness Studies*, *12*(8–9), 43–60.

Rickly, J. M. (2022). A review of authenticity research in tourism: Launching the annals of tourism research curated collection on authenticity. *Annals of Tourism Research*, *92*, 103349.

Rickly-Boyd, J. M. (2012). Authenticity aura: A Benjaminian approach to tourism. *Annals of Tourism Research*, *39*(1), 269–289.

Riva, G., Davide, F., & IJsselsteijn, W. A. (2003). Being there: The experience of presence in mediated environments. In G. Riva, F. Davide & W. A. Ijsselstein (Eds.), *Being there: Concepts, effects and measurement of user presence in synthetic environments* (pp. 4–16). Amsterdam: IOS Press.

Roberts, L. (1988). Conservationists in Panda-monium: US zoos and conservationists tangle over the propriety of borrowing pandas—an endangered species—and exploiting them in highly lucrative exhibits. *Science*, *241*(4865), 529–531.

Robertson, J. (2002). Reflexivity redux: A pithy polemic on" positionality". *Anthropological Quarterly*, *75*(4), 785–792.

Roche, D. G., O'Dea, R. E., Kerr, K. A., Rytwinski, T., Schuster, R., Nguyen, V. M., Young, N., Bennett, J. R., & Cooke, S. J. (2022). Closing the knowledge-action gap in conservation with open science. *Conservation Biology*, *36*(3), e13835.

Roosevelt, K., & Roosevelt, T. (1929). *Trailing the giant panda*. Charles Scribner's Sons. https://archive.org/details/trailinggiantpan0000kerm/page/n9/mode/2up

Rosa, R. M., Cavallari, D. C., & Vera-Silva, A. L. (2020). Pokécrustacea: The crustacean-inspired Pokémon. *Journal of Geek Studies*, *7*(2), 97–113.

Rose, N. (2009). *The politics of life itself*. Princeton University Press.

Rosenberg, K. R. (2021). The evolution of human infancy: Why it helps to be helpless. *Annual Review of Anthropology*, *50*, 423–440.

Rothfels, N. (2002). *Savages and beasts: The birth of the modern zoo*. JHU Press.

Ruedy, N. E., & Schweitzer, M. E. (2010). In the moment: The effect of mindfulness on ethical decision making. *Journal of Business Ethics*, *95*, 73–87.

Sandvoss, C. (2005). *Fans: The mirror of consumption*. Polity.

Santos, T. O., Correia, A., Biscaia, R., & Pegoraro, A. (2019). Examining fan engagement through social networking sites. *International Journal of Sports Marketing and Sponsorship*, *20*(1), 163–183.

Sasaki, M., Endo, H., Wiig, Ø., Derocher, A. E., Tsubota, T., Taru, H., Yamamoto, M., Arishima, K., Hayashi, Y., & Kitamura, N. (2005). Adaptation of the hindlimbs for climbing in bears. *Annals of Anatomy-Anatomischer Anzeiger*, *187*(2), 153–160.

Schaller, G. B. (1994). *The last panda*. University of Chicago Press.

Schebella, M. F., Weber, D., Schultz, L., & Weinstein, P. (2020). The nature of reality: Human stress recovery during exposure to biodiverse, multisensory virtual environments. *International Journal of Environmental Research and Public Health*, *17*(1), 56.

Schiebinger, L., & Swan, C. (2007). *Colonial botany: Science, commerce, and politics in the early modern world*. University of Pennsylvania Press.
Schultz, C. D. (2017). Proposing to your fans: Which brand post characteristics drive consumer engagement activities on social media brand pages? *Electronic Commerce Research and Applications*, *26*, 23–34.
Schussler, A. E. (2022). From Dasein to the Posthuman, Or toward a Rhizomatic-Coexistence-in-an-Open-World. *Journal of Posthuman Studies*, *6*(2), 151–167.
Scott, J. (1987). Zoo gets set for pandamania as 2 long-awaited visitors arrive: The Pandas are here! *Log Angeles Times*. Retrieved April 4, 2024, from https://www.latimes.com/archives/la-xpm-1987-07-24-me-3501-story.html
Sheldon, W. (1937). Notes on the giant panda. *Journal of Mammalogy*, *18*(1), 13–19.
Shih, S.-m. (2001). *The lure of the modern: Writing modernism in semicolonial China, 1917–1937* (Vol. 1). University of California Press.
Silverman, K. (1983). *The subject of semiotics*. Oxford University Press.
Slotkin, R. (2024). *A great disorder: National myth and the battle for America*. Harvard University Press.
Smith, A. (2013). Nations and nationalism in a global era. John Wiley & Sons.
Smith, M. K., & Diekmann, A. (2017). Tourism and wellbeing. *Annals of Tourism Research*, *66*, 1–13.
Songster, E. E. (2018). *Panda nation: The construction and conservation of China's modern icon*. Oxford University Press.
Spivak, G. C. (2003). Can the subaltern speak? *Die philosophin*, *14*(27), 42–58.
Stache, C. (2020). Conceptualising animal exploitation in capitalism: Getting terminology straight. *Capital & Class*, *44*(3), 401–421.
Stander, F. W., & De Beer, L. T. (2016). Towards engagement: A comparison of fan groups in the context of a major South African football club. *SA Journal of Industrial Psychology*, *42*(1), 1–10.
Stanfill, M. (2015). *Orienting fandom: The discursive production of sports and speculative media fandom in the Internet era*. University of Illinois at Urbana-Champaign.
Stanfill, M. (2019). *Exploiting fandom: How the media industry seeks to manipulate fans*. University of Iowa Press.
Steffen, W., Grinevald, J., Crutzen, P., & McNeill, J. (2011). The Anthropocene: Conceptual and historical perspectives. *Philosophical Transactions of the Royal Society A: Mathematical, Physical and Engineering Sciences*, *369*(1938), 842–867.
Stokes, G. (1986). How is nationalism related to capitalism? A review article. *Comparative Studies in Society and History*, *28*(3), 591–598.
TallBear, K. (2000). Shepard Krech's the ecological Indian: One Indian's perspective. *International Institute for Indigenous Resource Management [IIIRM] Publications*, *30*, 1–5.
Tam, G. A. (2023). "Our roots are the same": Hegemony and power in narratives of Chinese linguistic antiquity, 1900–1949. *Comparative Studies in Society and History*, *65*(1), 27–52.
Taylor, C. (2014). Biopower. In *Michel Foucault* (pp. 41–54). Routledge.
Terranova, T. (2012). Free labor. In *Digital labor* (pp. 33–57). Routledge.
Terrone, A. (2014). The earth as a treasure in Tibetan Buddhism: Visionary revelation and its interactions with the environment. *Journal for the Study of Religion, Nature and Culture*, *8*(4), 460–482.
Times, G. (2023). *Netizens celebrate birthday of giant panda top stars in Chengdu and Beijing*. Retrieved May 21, 2024, from https://www.globaltimes.cn/page/202307/1293694.shtml
To, S. (2013). Understanding sheng nu ("leftover women"): The phenomenon of late marriage among Chinese professional women. *Symbolic Interaction*, *36*(1), 1–20.
Tooby, J., & Cosmides, L. (2001). Does beauty build adapted minds? Toward an evolutionary theory of aesthetics, fiction, and the arts. *SubStance*, *30*(1), 6–27.
Torres, B. (2007). *Making a killing: The political economy of animal rights*. AK Press.
Trail, G. T., Anderson, D. F., & Lee, D. (2017). A longitudinal study of team-fan role identity on self-reported attendance behavior and future intentions. *Journal of Amateur Sport*, *3*(1).

Tsing, A. (2013). More-than-human sociality: A call for critical description. In *Anthropology and nature* (pp. 27-42). Routledge.

Urry, J., & Larsen, J. (2011). *The tourist gaze 3.0* (3rd ed.). Sage.

Van Dooren, T., & Rose, D. B. (2012). Storied-places in a multispecies city. *Humanimalia, 3*(2), 1–27.

Van Krieken, R. (2018). *Celebrity society: The struggle for attention*. Routledge.

VCG. (2024). *Panda frenzy*. Retrieved June 28, 2024, from https://www.globaltimes.cn/page/202401/1304824.shtml

Veblen, T. (2007). *The theory of the leisure class*. Oxford University Press.

Villamagna, A. M., Mogollón, B., & Angermeier, P. L. (2017). Inequity in ecosystem service delivery: Socioeconomic gaps in the public-private conservation network. *Ecology and Society*, *22*(1).

Wadiwel, D. J., & Fanon, F. (2020). Thinking "critically" about animals after colonialism. In K. S. Montford, & C. Taylor (Eds.), *Colonialism and animality: Anti-colonial perspectives in critical animal studies*. Routledge.

Walsh, D. (1998). *Doing ethnography* (Vol. 2012). Sage.

Wang, F. (2023). Chinese eagerly await return of panda from US zoo. *BBC*. https://www.bbc.com/news/world-asia-china-65229899

Wang, H. (2019). 王怀林 | 《寻找康巴》 第一章西部民族在迁徙（八）：千年土司——中央治下的"土皇帝"们. Retrieved March 20, 2023, from https://scsqw.cn/whzh/wxz

Wang, N. (1999). Rethinking authenticity in tourism experience. *Annals of Tourism Research*, *26*(2), 349–370.

Wang, X. (2016). *Social media in industrial China*. UCL Press.

Wang, Z. (2014). *Never forget national humiliation: Historical memory in Chinese politics and foreign relations*. Columbia University Press.

Wangjie, G., Lou, Y., Song, K., Li, X., Fang, Y., Hou, J., & Sun, Y. (2024). Human–wildlife coexistence on the Qinghai–Tibet Plateau: The correlation between Tibetan Buddhist monasteries and avian community composition. *People and Nature*.

Weiss, J. C., & Wallace, J. L. (2021). Domestic politics, China's rise, and the future of the liberal international order. *International Organization*, *75*(2), 635–664.

Williams, R. (2015). *Post-object fandom: Television, identity and self-narrative*. Bloomsbury Publishing.

Wilson, E. H. (1913). *A naturalist in Western China, with vasculum, camera, and gun: Being some account of eleven years' travel, exploration, and observation in the more remote parts of the flowery kingdom* (Vol. 1). Doubleday, Page & Company.

Wu, F. H. (2002). The best "chink" food: Dog eating and the dilemma of diversity. *Gastronomica*, *2*(2), 38–45.

WWF. (2022). *Giant pandas: Living proof that conservation works*. Retrieved January 6, 2022, from https://www.wwf.org.uk/learn/wildlife/giant-pandas

Xing, Y. (2010). *China's panda diplomacy: The power of being cute*. University of Southern California.

Yin, J., Gao, Y., & Xu, H. (2014). Survey and analysis of consumers' behaviour of waste mobile phone recycling in China. *Journal of Cleaner Production*, *65*, 517–525.

Youatt, R. (2008). Counting species: Biopower and the global biodiversity census. *Environmental Values*, *17*(3), 393–417.

Zhang, X., & Chen, C. (2007). Zàngzú chuántǒng jùluò xíngtài yǔ Zàngchuán Fójiào shìjièguān [Traditional settlement patterns of Tibetan communities and the worldview of Tibetan Buddhism]. Zōngjiào Xuéyánjiū [*Journal of Religious Studies*], *2*, 202–206.

Zhang, Z., & Zhou, J. (2020). Cognitive and neurological mechanisms of cuteness perception: A new perspective on moral education. *Mind, Brain, and Education*, *14*(3), 209–219.

Zoo, B. (2022). *Giant panda*. Retrieved October 18, 2022, from https://www.zoo-berlin.de/en/artenschutz/weltweit/giant-panda

Zsok, F., Haucke, M., De Wit, C. Y., & Barelds, D. P. (2017). What kind of love is love at first sight? An empirical investigation. *Personal Relationships*, *24*(4), 869–885.

何芬奇, & 孙前. (2009). 对大熊猫古名问题的思考(上). 大自然, *4*, 14.

侯光, & 蒋永志. (1991). 嘉绒藏族眼中的自然美. 文史杂志, *3*, 27–29.

孙前, 何芬奇, & 孙震. (2008). 大熊猫古名研究. 动物分类学报, *33*(4), 819–826.

张雪梅, & 陈昌文. 藏族传统聚落形态与藏传佛教的世界观. 宗教学研究, *2*, 202–206.

林鸿荣, & 林林. (2005). 也谈大熊猫之今古称谓. 北京林业大学学报: 社会科学版, *4*(1), 49–52.

胡锦矗. (1981). 大熊猫的历史记载. 南充师院学报(自然科学版), *3*, 7–14.

INDEX

Note: Numbers in **bold** indicate a table.

For Product Safety Concerns and Information please contact our EU
representative GPSR@taylorandfrancis.com
Taylor & Francis Verlag GmbH, Kaufingerstraße 24, 80331 München, Germany

www.ingramcontent.com/pod-product-compliance
Ingram Content Group UK Ltd.
Pitfield, Milton Keynes, MK11 3LW, UK
UKHW020140200926
475702UK00005B/838
* 9 7 8 1 0 3 2 8 4 1 0 3 8 *